大熊猫不仅是福州的，也是全国全世界人民的
共同财富。

——习近平

Giant pandas are not only a treasure of
Fuzhou, but also a treasure of the world.

—Xi Jinping

Basi: The Ambassador of Peace

和平使者
熊猫巴斯

高富华

编著

[新西兰] Donna Jiang （蒋梓青）
[新西兰] Rocky Jiang （蒋梓恒）

陈思嘉

译

海峡出版发行集团 THE STRAITS PUBLISHING & DISTRIBUTING GROUP | 福建人民出版社 FUJIAN PEOPLE'S PUBLISHING HOUSE

目录

Contents

第一章　巴斯向世界人民问好

Chapter 1　Basi says hello to the world

纽约时报广场。
New York's Times Square.

2015年9月22日，3D动画片《巴斯向世界人民问好》在美国纽约时报广场"中国屏"上滚动播放。《巴斯向世界人民问好》的主角正是有着"世界和平大使"之称的大熊猫巴斯。这是巴斯阔别28年之后，再次"重返"美国。

3D动画片播出期间，正值中国国家主席习近平赴美参加联合国成立70周年暨世界反法西斯战争胜利70周年纪念活动。3D动画片播出后，吸引了众多美国人和来自世界各地的游客驻足观看。

巴斯来自中国，秉承中国爱好和平的愿望，向国际社会传达了中国对世界和平的渴望、对美好和平生活的向往。

3D动画片《巴斯向世界人民问好》在纽约时报广场"中国屏"上滚动播出。
The 3D cartoon "Basi says hello to the world" was displayed on the "China Screen" in New York's Times Square.

自古以来，大熊猫就是和平的象征、友谊的使者。

在古代，人们称大熊猫为"貔貅""食铁兽""驺虞"……大熊猫主要吃竹子，性情温顺，很少主动攻击其他动物或人，有"义兽"之美誉而成为和平象征。两国交战时，若一方举起画着熊猫的旗帜，要求和平友好，对阵双方就会停止厮杀，战斗就会停息下来。

公元685年，女皇武则天送给日本天武天皇两只大熊猫，这是大熊猫第一次作为"友好使者"出国。此后不断有大熊猫担当起和平友谊的使者，增进了中国和世界各国人民的感情。

On the 22nd of September 2015, the 3D cartoon "Basi says hello to the world" was displayed. On the "China Screen" in New York's Times Square. The leading protagonist, a giant panda, is known as the World Ambassador of Peace, and this was her return to the United States of America after 28 years.

At the time of broadcast, Chinese President Xi Jinping was present in the United States of America to commemorate the 70th anniversary of both the World Anti-Fascism War and the founding of the United Nations. While displayed, the 3D animation attracted the attention of tourists from all walks of life.

Basi is from China and is a symbol of China's love of peace. Basi conveys to the international community China's deep desire for peace, and China's commitment to building beautiful, peaceful lives.

Pandas have always been ambassadors for peace and friendship.

Giant pandas primarily eat bamboo and are of mild temperament. It is extremely rare for them to actively attack other animals and people, and have been called "peaceful beast" in the past. In the case of war, a flag with a panda on it has the same meaning as that of a white flag—to request a ceasefire or truce.

The first recorded instance of pandas being used as ambassadors traces back to 685 AD, when Empress Wu Zetian sent two pandas as gifts of peace to Emperor Temmu of Japan. Since then, pandas have continued to be important ambassadors, strengthening ties between China and other countries.

2015年9月22日，3D动画片《巴斯向世界人民问好》在纽约时报广场"中国屏"上滚动播出，吸引了众多纽约市民驻足观看，并在美国华人社会引起轰动。

On the 22nd of September 2015, the 3D cartoon "Basi says hello to the world" was displayed on the "China Screen" in New York's Times Square, causing a stir among the American Chinese community as New York City residents stopped to watch the broadcast.

可爱的大熊猫
嘴里吃着
葡萄紅蘿
蔔眼
底看看
蘋果著红
好忙忙
好蜻蜓
蝴蝶
都飛來取
笑他
了呀

璦容
淑德

巴斯精彩瞬间。
Wonderful moments of Basi's.

第二章　传奇熊猫冰河重生

Chapter 2　A panda's survival after falling into a frozen river

幼年的巴斯在树上。
Young Basi in the tree.

巴斯落水

1980年，大熊猫巴斯出生在四川省宝兴县的夹金山中，在大山的怀抱里，她度过了快乐的童年。然而好景不长，1984年春天，一场灾难席卷而来，夹金山上的箭竹大面积开花，大熊猫们一下就断了粮。

1984年2月22日下午，饥肠辘辘的巴斯摇摇晃晃地往山脚走。

开过花的竹子都枯死了。巴斯在山上已经转悠了好几天，找不到一点儿吃的。困守大山，等待巴斯的就是死亡。

巴斯望了望山腰处的炊烟，嗅了嗅，空气中飘来了一股股肉香。巴斯相信人类朋友是会帮助她度过这段困难时期的。

巴斯试探着往山腰爬去，然而一条小河挡住了她的去路。巴斯打量了一下，小河的水并不深，巴斯决定趟过去。谁知巴斯刚踏进漂浮着冰凌的小河，一下就滑倒在水中。水流很急，巴斯拼命挣扎也无法站立起来，随后巴斯被卡在了石缝中……

李兴玉实地讲述当年从冰河中救起巴斯的过程。
Li Xingyu was telling the story of rescuing Basi from the icy river.

由于大熊猫依赖竹子为生，竹子周期性开花死亡，对环境中竹种单一地区的大熊猫生存是致命的威胁。20世纪70年代和80年代，四川岷山、邛崃山竹大面积开花就曾造成大熊猫种群数量明显下降。

1980年，为了拯救大熊猫这一濒危物种，世界野生生物基金会（现称世界自然基金会）与我国政府达成为拯救大熊猫而进行国际募捐运动和制定保护熊猫计划的协议，并派出以乔治·夏勒为首的科学家到四川与中国科研人员共同探讨执行保护大熊猫的计划。

Due to the periodic blooming cycle of bamboo, pandas in specific areas where only one type of bamboo grows are endangered. In the 1970s and 1980s, the blooming of large patches of bamboo in the mountain areas of Minshan and Qionglaishan in Sichuan caused a significant decline in the population of giant pandas.

In order to save the endangered giant pandas, the World Wide Fund for Nature (now called World Wildlife Fund) and the Chinese government reached an agreement for the international fundraising campaign to save giant pandas and to raise awareness of giant panda protection. The organization sent a team of scientists led by G·B·Schaller to Sichuan to discuss and plan the panda preservation initiative with the Chinese scientists and preservation workers.

Basi in the water

Basi was born in 1980, on Jiajin Mountain, Baoxing County, Sichuan Province. Hidden in the embrace of the mountain, she enjoyed a happy childhood. However, this was not to last, as in the spring of 1984, the bamboo trees on Jiajin Mountain blossomed all at once. This meant that after their flowering, the bamboo trees would die, leaving a shortage of food for giant pandas.

On the afternoon of 22nd February 1984, Basi stumbled down to the base of the mountain.

The bamboo which had blossomed had now died. Basi had been foraging for many days, to no success. Staying in the mountains could only mean death.

Basi could see the smoke in the distance. Raising her snout to the air, she sniffed. The breeze carried the smell of delicious food, and Basi believed that humans would help her survive through the difficult days.

As Basi made her way towards the source of that smell, she found that a small river blocked her path. Sizing up the river, Basi decided that it wasn't so deep, and decided to attempt to cross. As soon as she stepped into the icy river, however, she fell into the water. The water was moving rapidly, and Basi was unable to stand. She found herself trapped between some rocks in the river, held by the current...

巴斯出生地——夹金山。
The birthplace of Basi: Jiajin Mountain.

冰河获救

就在巴斯生命垂危之际，她依稀记得有人向她奔跑过来。

原来，当地的农民李兴玉正好上山砍柴回来，跟李兴玉一起的还有邻居张天玉的儿子石家明。当他们背着柴火回家途经"巴斯沟"时，几乎都看到巴斯在激流中挣扎的身影。

"大熊猫！大熊猫落水啦！"石家明叫嚷了起来。

李兴玉忙把背上的柴火扔在地上，一头冲到了河边，刚好看到巴斯被冲到河中央，卡在了两块大石头中间。听到呼唤声，巴斯费尽力气，好不容易才抬起了头，看了李兴玉他们一眼，又垂下了脑袋，随后陷入了昏迷。

李兴玉让石家明跑回去喊人帮忙。她解下了捆柴的绳子，一头拴在自己身上，另一头正准备拴在树上，石家明和他的母亲张天玉赶来了。

"你们拉住绳头，万一大熊猫倒了，你们不要松绳子。"李兴玉边说边往河里走。夹金山终年积雪，急流中挟带着冰凌，冰冷刺骨的河水冻得李兴玉浑身发抖，尽管如此，李兴玉还是使尽力气，一步一步地向巴斯走了过来。

巴斯早已被冻僵了。李兴玉轻轻地把巴斯从冰水中抱起来，在张天玉母子的牵引下，李兴玉抱着巴斯艰难地走上河岸。

年幼的巴斯。
Young Basi.

由于在冰河中待的时间太久，当时的巴斯已经没有知觉。李兴玉急了，她伸出手在巴斯的鼻子下面摸了摸，发现巴斯还有一丝微弱的气息。

李兴玉早已忘记了寒冷，她顾不上脱下被冰河打湿的衣服就解开衣服，把巴斯抱在心窝上，用自己的身体为巴斯取暖。张天玉点燃了柴火后，李兴玉把巴斯放在火堆边取暖。

李兴玉一口气跑回家，换了衣服后，又飞快地来到了巴斯身边。她还拿来了红糖、玉米面等食物，为巴斯准备了一顿大餐。

Rescue from the icy river

As Basi's situation became precarious, two human figures ran towards her.

It turned out that a local farmer by the name of Li Xingyu had just been returning from gathering firewood. With her was her neighbour's son Shi Jiaming. It was only by chance that they noticed Basi's struggling form in the river.

Throwing the firewood aside, Li Xingyu rushed to the riverside. There, she saw Basi in the middle of the river, swept up between two rocks. Hearing the voices, Basi managed to lift her head. Seeing the men on the riverbank, she fainted.

Li Xingyu instructed Shi Jiaming to fetch help as quickly as possible. Li Xingyu herself untied the rope holding the firewood together, tying one end to herself. Just as she was going to tie the other end to a nearby tree, Shi Jiaming and his mother Zhang Tianyu arrived.

人工喂养大熊猫幼崽。
A keeper was feeding panda cubs.

"You guys hold onto this end of the rope. If the giant panda happens to fall, hold on as tight as you can." she instructed them. As she was doing so, she began wading into the river. The cold, biting rapids left Li Xingyu shivering, but tirelessly, she moved towards Basi.

Basi had long been paralyzed by the cold. Li Xingyu gently picked Basi up, and with the help of Shi Jiaming and his mother, walked towards the riverbank with obvious difficulty.

Having been in the freezing river for too long, Basi had lost consciousness. Feeling worried, Li Xingyu felt under Basi's snout. Basi was still breathing, although very weakly.

Li Xingyu didn't, in her exhilaration, care about the cold. Removing her clothing, she cradled Basi in her arms, keeping Basi warm. Only once Zhang Tianyu had lit a fire did Li Xingyu place Basi gently by the fire, keeping her warm. Li Xingyu then returned home, changed her wet clothes, and returned to Basi's side. She also brought brown sugar, cornmeal and other foods, preparing a meal for Basi.

"小宝贝，你怎么还不醒来？你看我已经给你拿来食物了。"李兴玉边喃喃自语边轻轻地为巴斯梳理毛发。随着火堆温度的上升，在李兴玉的深情呼唤下，巴斯慢慢地苏醒了过来，瞪着一双眼睛骨碌碌地看着李兴玉等人。

四川省蜂桶寨国家级自然保护区管理局干部高华康等人闻讯赶了过来。他们把巴斯送到位于夹金山下的四川省蜂桶寨国家级自然保护区管理局大水沟管护站，让巴斯接受人工饲养。临行前，李兴玉就以发现这只大熊猫的地点"巴斯沟"，为她取名叫巴斯。

迎着第一缕晨曦，巴斯被送入与"巴斯沟"相距不远的大水沟大熊猫管护站。在那里，医生对她进行了全面的身体检查，发现她是一只雌性大熊猫，虽然身体虚弱，但好在没有什么毛病，并确定巴斯的年龄在4岁左右。

巴斯早期档案。
The early file of Basi's.

大熊猫的祖先在大约800万年前就已经出现在地球上。在距今50万—70万年时，大熊猫种群进入鼎盛时期。在更新世中晚期，秦岭及其以南山脉出现大面积冰川等自然环境的剧烈变化，特别是在距今约18,000年前的第四纪冰川期之后，大熊猫－剑齿象动物群衰落，大部分动物灭绝，仅留下无数化石表明它们曾经存在。唯独大熊猫在这次劫难中幸存了下来，它们改变了自己的食性来适应环境的变化，从食肉变为食竹。过去的一万年，人类文明的发展不断地侵蚀大熊猫的自然领地，迫使大熊猫退缩隐居于青藏高原东部边缘的高山深谷之中。

The ancestors of giant pandas have been on Earth for the last eight million years. Around 500,000 to 700,000 years ago, the population of pandas related species was at its highest. Towards the end of the Pleistocene, dramatic changes in the climate occurred around the Qinling Mountains and other mountain ranges to the south. Around 18,000 years ago, the population of animals such as the giant pandas and woollen mammoths decreased severely. Many other animals went extinct, leaving only fossils behind to indicate their existence. The giant pandas survived this period, adapting by becoming predominantly herbivorous. Over the last 10,000 years, however, humankind has continually encroached upon giant pandas' territory, forcing giant pandas to move towards the valleys on the eastern side of the Tibetan Plateau.

"Come dear. Wake up now. I've brought you food," she whispered to Basi, while stroking her fur. Basi slowly awoke in this affectionate calling as the temperature of the fire rose, staring at the humans around her.

Gao Huakang soon arrived with others from Fengtongzhai Nature Reserve Management Bureau of Sichuan. They then transported Basi to Dashuigou Panda Conservation Centre of Fengtongzhai Nature Reserve Management Bureau located at the base of Jiajin Mountain, where she was taken under the care of humans. Just before the departure of Basi, Li Xingyu named her Basi as the place that she had been stuck was known as Basi Valley.

The next morning, Basi arrived at Dashuigou Panda Conservation Centre. There, a physical examination was conducted. It was there that they discovered that she was female, and while she was extremely weak from the previous days, she was relatively healthy otherwise. They also found that she was around four years old.

2005年，李兴玉受邀参加巴斯25周岁生日庆典，并为出席庆典活动的嘉宾讲述抢救巴斯的过程。

In 2005, Li Xingyu was invited to Basi's 25th birthday celebration, and she told the guests in attendance about the process of saving Basi.

國際揚名熊貓巴斯表演精彩不僅深入虜建人心群蝶也飛舞左右來湊熱鬧　瑗谷淑德

巴斯精彩瞬间。
Wonderful moments of Basi's.

第三章　福州大梦山上演亲情
Chapter 3　Fondness upon Dameng Hill of Fuzhou

巴斯和她第二个故乡的"家人"。
Basi and her family of her second hometown.

四川"接亲"

1977年，海峡（福州）熊猫世界从甘肃引进了两只抢救的大熊猫。20世纪80年代初，海峡（福州）熊猫世界主任陈玉村带领他的团队，通过驯化研究，改变了对大熊猫进行体检或治疗时必须采取麻醉和捆绑的传统做法，在与大熊猫建立充分信任的基础上，为我国首次获得大熊猫无干扰生理数据，创造了世界纪录。一时间，海峡（福州）熊猫世界享誉海内外。陈玉村主任抓住这个机会，在时任福建省委书记项南的支持下，向当时的国家林业部提出再增加两只大熊猫给海峡（福州）熊猫世界的申请。

1984年4月，原国家林业部正式下文，批准巴斯和元元两只大熊猫给海峡（福州）熊猫世界。

拿着"批文"并不等于就能领走大熊猫。海峡（福州）熊猫世界一位副主任两次到四川都是乘兴而来，扫兴而归。相貌俊美、聪明伶俐又将到生育年龄的"美女"巴斯，早被四川省卧龙自然保护区管理局大熊猫研究中心看中了。卧龙大熊猫研究中心的工作人员千方百计要留下巴斯。当时，他们正在开展大熊猫人工繁育科研，希望她能养育出漂亮的大熊猫幼仔。

两次索要不成之后，陈玉村只身前往四川。他当时只有一个念头：无论如何也要把巴斯带回来。以前，陈玉村多次到过卧龙。卧龙大熊猫研究中心上至领导，下至工作人员，陈玉村认识的人不少，可是这次仿佛到了一个陌生的地方，熟人一个也没见到，一打听，说是都出差了。

巴斯和陈玉村。
Basi and Chen Yucun.

Welcoming Basi from Sichuan

In 1977, Strait（Fuzhou）Panda World introduced two rescued giant pandas from Gansu Province. During the early 1980s, Chen Yucun, director of the panda world, led his team to change the traditional practices of bundling up the pandas and putting them under anaesthesia for physical examinations. Chen Yucun managed to establish a relationship with pandas in order to gain their trust. For the very first time, Chen Yucun and his team were able to collect biological information about the pandas without being intrusive to their wellbeing. For now, Strait（Fuzhou）Panda World is renowned locally and internationally. Chen Yucun has seized this opportunity to submit an application to the former State Ministry of Forestry for the addition of two more pandas to Strait（Fuzhou）Panda World, with the support of Xiang Nan, Secretary of CPC'S Fujian Committee.

In April 1984, the former State Ministry of Forestry formally approved the application and permitted the transfer of Basi and Yuanyuan to Strait（Fuzhou）Panda World. However, having the approval did not allow someone to transfer the pandas immediately. Twice, the deputy director of Strait（Fuzhou）Panda World arrived in Sichuan, only to leave alone. The beautiful and intelligent Basi would soon reach the age for reproduction, and had long been favoured by the Giant Panda Research Centre at Wolong Nature Reserve Administration in Sichuan Province. The staff of the centre were trying everything to make sure that Basi could stay. At the time, they were carrying out research into the artificial breeding of giant pandas, in hopes, that she would produce beautiful panda cubs.

After the two unsuccessful attempts to transfer Basi to Strait（Fuzhou）Panda World, Chen Yucun went to Sichuan alone, focusing on bringing Basi back no matter what the effort. In the past, Chen Yucun had visited the center many times, and was well acquainted with many staff members. However, this time, he was unable to find a familiar face as he was told that they had gone on a business trip.

巴斯和陈玉村。
Basi and Chen Yucun.

陈玉村买好干粮，一屁股就坐在了时任卧龙自然保护区党委书记、管理局局长赖炳辉的家门口，来个"守株待兔"。凌晨一点，打着哈欠的赖炳辉终于回家了。还没进门，就被守候多时的陈玉村"逮"住。赖炳辉只得请他进门商量。历经几次倾谈，卧龙自然保护区管理局和卧龙大熊猫研究中心的领导们终于同意把巴斯让给福州。

　　好事多磨！卧龙大熊猫研究中心的饲养员与巴斯有了感情，饲养员以科研需要为由，就是不愿意放走巴斯。陈玉村无奈之下，只得"祭"出撒手锏："你们满山的大熊猫，怎么就不把巴斯让给福州呢？好！我这就上北京，找董智勇副部长和你们说。"

　　饲养员只好含泪送走巴斯。

幼年的巴斯。
Young Basi.

　　Chen Yucun bought some dry food and sat down outside the house of Lai Binghui, director of Wolong Nature Reserve Administration. When Lai Binghui, who finally yawned at one o'clock in the morning, prepared to return home, he found Chen Yucun waiting. Because Chen Yucun had been waiting for such a long period, Lai Binghui had to invite him in. After talking with Chen Yucun for several times, the leaders of Wolong Nature Reserve Administration and Wolong Giant Panda Research Centre finally agreed to give Basi to Fuzhou.

　　Good things come to those who wait and work hard. The keeper at the research centre, however, had formed a relationship with Basi, and was unwilling to part ways with her. Having no other way to persuade the keeper, Chen said, "You have so many pandas in the mountains, yet you cannot give up one for Fuzhou! Fine, I will go to Beijing to ask the Deputy Minister Dong Zhiyong to come to talk to you." The keeper had no other choice but to tearfully send Basi away.

福州安家

1984年5月，巴斯终于来到了福州。摆在陈玉村面前的难题就是谁能担任巴斯的饲养员。从大山到都市，面对陌生的世界，巴斯有些不习惯。野性十足的她根本没有一点"大家闺秀"的样子，生猛得很，一言不合，就伸出蒲团似的巴掌扫过来，稍微细一点的钢筋围栏都要被她扫弯。

巴斯第一次来到陌生的城市，加上入夏后，福州天气变得潮湿闷热，她变得坐立不安、无精打采，开始吃不下食物。为了消除巴斯与新环境的心理隔阂，陈玉村想到了一个好办法——让饲养员和她"交朋友"，并和巴斯零距离接触。

海峡（福州）熊猫世界面向社会公开招聘饲养员。不久，一个19岁的江南美女陈小玲来到了巴斯身边。此后10年里，陈小玲几乎整日与巴斯在一起。陈小玲用她的耐心、细心和爱心，与巴斯日夜相伴，并最终和巴斯成为亲如家人的好朋友。

巴斯和陈小玲。
Basi and Chen Xiaoling.

Settling in Fuzhou

In May 1984, Basi finally arrived in Fuzhou. The challenge then was, who would be the keeper of Basi. Moving from the mountainous areas to the city, Basi needed to adapt. She was not used to being spectated by so many and she was very fierce. Anything that made her uneasy resulted in a swipe of her paw. Even a fine steel fence is no match for her.

Initially, when Basi arrived at a foreign city to her, Fuzhou's weather became hot and humid after summer came. She became restless and slouched, losing her appetite. In order to help Basi settle down and become accustomed to her new environment, Chen Yucun decided to let the keeper create a friendship and have direct contact with her.

Strait (Fuzhou) Panda World openly recruited keepers from the community. Not long after, a 19-year-old *Jiangnan* (regions south of the Yangtze River) girl, Chen Xiaoling, came to Basi. In the ten years after their first encounter, Chen Xiaoling was with Basi almost every day. Her patience, carefulness and love for Basi day and night made their relationship alike to that of kin.

巴斯在笼舍内。
Basi in her enclosure.

音乐和体育天赋

巴斯大概没想到自己会跟音乐和体育联系到一起。这还要从陈玉村主任想方设法缓解她因不适应福州的生活而焦躁不安说起。陈玉村主任除了让饲养员和巴斯交朋友和必要的饮食调整外，还对她的居住环境进行了降温、除潮等，甚至还为她在福州鼓岭专门建了避暑用的熊猫山庄。在缓解巴斯不适的过程中，陈玉村先是让巴斯听音乐。在听音乐的训练中，陈玉村感觉巴斯入了神，她对音乐有奇妙的反应。当时，第23届奥运会在美国洛杉矶举行，福建籍运动员许海峰拿下了中国首枚奥运金牌，中国大地刮起了一阵奥运风。陈玉村受到启发，决定让饲养员陈小玲带着巴斯跟着音乐做体能运动，通过运动增强巴斯的体质，加强与人类的情感沟通，消除恐惧，亲近人类，进而为科研医疗服务。

意外之喜就这样发生了！巴斯天赋极佳，居然把投篮、举重等动作做得有模有样。她也许天生就是个运动健儿，听到优美的音乐，她就会张大耳朵。她爱上了音乐，爱上了运动，也缓解了原来焦躁不安的情绪。她在音乐和运动的影响下变得温顺起来，也逐渐喜欢与人相处，与人亲近了……

巴斯拉二胡。
Basi was playing the Chinese musical instrument *erhu*.

Music and sporting talents

Basi perhaps never thought she would be associated with music and sports. This started when Chen Yucun tried to ease her anxiety while she was adjusting to her new life in Fuzhou. In addition to letting Basi form a friendship with her keeper, dietary adjustments were made when necessary, and in her living space, the temperature was reduced and the area protected from moisture. A special panda house was also built specifically as a summer resort at Kuliang, Fuzhou for Basi. Chen Yucun decided to play music to soothe Basi's discomfort which she would later develop an affinity for, prompting Chen to describe Basi being enthralled by the music due to her wonderful response to it. At the time, it was the 23rd Olympic Games held in Los Angeles. Fujian-born athlete, Xu Haifeng, won China's first Olympic gold medal, and then China was caught up in a whirlwind of excitement. Feeling inspired, Chen Yucun decided to let Basi's keeper, Chen Xiaoling, conduct exercises with Basi to music to improve her health and enhance her emotional communication with humans, eliminate her fears and to allow her to trust humans. Doing all these will work for the further scientific research and medical treatment on Basi.

Joy can often be found in serendipitous situations. Basi had an extraordinary talent being able to shoot hoops and lift weights. She had a natural talent to be a sports athlete. When she heard music, she would open her ears. Basi fell in love with music, fell in love with sports, but also eased her anxiety. She calmed down under the influence of music and sports. She allowed people to get close to her and got along with them...

巴斯进行体育运动。
Basi was doing sports.

在动感的音乐中，陈小玲与巴斯亲密相处，其乐融融。要知道，巴斯和其他熊科动物一样有攻击性，性情大多凶猛，但陈小玲与巴斯相处的时间却是温馨的。陈小玲常常一边轻轻地梳理巴斯的毛发，一边与巴斯小声地交流着。

陈小玲与巴斯亲近后，巴斯慢慢能听从她的口令。陈小玲一个口令，一个指挥动作，巴斯都能迅速反应——让她躺下，她就躺下，让她仰卧，她就仰卧。陈小玲与巴斯成了亲密相处的好朋友，也为巴斯的健康检查打下基础！

也许是无意的训练，也许巴斯只是想与饲养员有密切的接触，有亲密的沟通，也许只是简单的躺下，翻滚的动作，但在这享受的过程中，巴斯完成了一个又一个美妙的动作，完成了一次又一次的飞跃。

经过一段时间的训练，巴斯学会了举重、抱娃娃、坐摇椅、直立行走等。两年后，巴斯竟学会了投篮、骑自行车、套环等十几种技艺！

巴斯投篮球。
Basi was playing basketball.

巴斯骑自行车。
Basi was riding a bicycle.

大熊猫名字中有"猫"，但实际属于熊科。它们胖嘟嘟的身体和内八字慢吞吞的行走方式，让人觉得"笨"得可爱，但事实上它们非常灵活，可以跑得很快，特别擅长爬树，能爬上高20米以上的树。大熊猫前肢臂力强，很粗的竹子，一掰就断；它们还喜欢把身体摆成各种各样的姿势。

In Chinese, the characters for "pandas" include the character for "cats", but in reality, they belong to the bear family. With their chubby body and their slow yet swift walking style, people often view them as cute and perhaps a little clumsy. They are, however, quite flexible and can run very fast. They are particularly good at climbing trees and can climb more than 20 meters high. Pandas have strong forearms, even the thickest bamboo can be broken into two by these giant bears. They also like to position their bodies in different poses.

巴斯玩晃板、套环。
Basi was playing on a rocking board and catching rings.

巴斯玩摇摇马。
Basi was rocking a horse.

Music evoked the strongest of sentiments and through this, Basi became very fond of Chen Xiaoling and they formed a close relationship. It should be noted that Basi was as aggressive as any other bear, with a fierce temperament, but to Chen Xiaoling, Basi was kind and warm. Often, Chen Xiaoling combed Basi's hair and communicated quietly with her.

As this relationship developed, Basi began understanding Chen Xiaoling's instructions and would act upon her commands immediately whether it was to lie down or sit back. Chen Xiaoling and Basi had become good friends which made Basi's physical examinations easier.

Perhaps it was unintentional conditioning, maybe Basi just wanted to have that contact with her keeper or maybe she simply enjoyed lying down and rolling around, but in this process, Basi was able to overcome different challenges one by one.

After a period of training, Basi learned how to lift weights, hold dolls, sit on a rocking chair and walk upright. Two years later, Basi learned as many as a dozen new techniques!

零距离获得大熊猫无干扰生理数据

为了让巴斯能安静地接受科研人员对她的身体进行各种检查，使科研工作有新的突破，陈小玲和海峡（福州）熊猫世界其他饲养员们开始对巴斯进行集训。经过长时间的探索，陈小玲他们发现，晨昏和夜间是最好的训练时机。巴斯在这些时间段活动频繁，此时她处于饥饿状态，而且周围的环境也很安静，外界干扰少。陈小玲每天起早贪黑，抓住这些时机，对巴斯进行耐心的训练。

在陈小玲"姐姐"的口令和指挥下，巴斯会静静地躺下，有时一躺就是一两个小时。虽然有些难受，但巴斯还是坚持下来，耐心地接受科研人员对她进行心脏听诊、测体温、量血压、B超检查、采血和输液等各项检查和治疗，为我国获得珍贵的大熊猫无干扰生理数据。

饲养员们在训练中发现，巴斯对酒精的气味非常敏感。为了减少酒精对巴斯的刺激，陈小玲想了很多办法。为了照顾巴斯敏感的情绪，给巴斯涂抹伤口时，陈小玲先是用纯净水，然后逐渐增加酒精浓度，尽量做到不引起巴斯的不适。

不麻醉情况下，巴斯接受采血检查。
Basi was accepting blood collection without being tranquilized.

Facilitating access to physiological data collection of giant pandas

陈玉村给巴斯注射。
Chen Yucun was administering an injection on Basi.

陈玉村为巴斯进行检查。
Chen Yucun was initiating the examinations on Basi.

In order to enable Basi to accept the scientists carrying out various body checks and to make new breakthroughs in their research, Chen Xiaoling and others of Strait (Fuzhou) Panda World began to train Basi. After a long period of observation, Chen Xiaoling found that early mornings and nights were the most opportune for training. Basi was active during these periods and hungry. Her surrounding environment was also very quiet with little external interruptions. Chen Xiaoling got up early every day and seized the opportunity to conduct training patiently on Basi.

Under the instructions of her keeper, Basi would lie down obediently, sometimes for an hour or two. Although it was uncomfortable, Basi persisted, and patiently allowed the examinations and treatments to be conducted—including those for cardiac auscultation, temperature measurement, blood pressure measurement, B-ultrasound, blood collection and fluid infusion. Through this, China was able to collect vital information about giant pandas without interruptions.

The keepers found that Basi was very sensitive to the smell of alcohol when they conducted training. To overcome this, Chen Xiaoling tried many ideas to help Basi. When treating Basi's wounds, Chen Xiaoling would initially use pure water before slowly increasing the alcohol concentration so that Basi would not be uncomfortable.

巴斯成就两岸佳缘

1995年，陈小玲"姐姐"在陪伴巴斯十年后才依依不舍地离开，嫁到台湾海峡的另一边——台湾。

说起"姐姐"的海峡姻缘，巴斯还算是"月老"呢！

陈小玲"姐姐"陪伴了巴斯十年，除了在生活上照顾巴斯，还教会巴斯各种体育"特技"。从美国圣地亚哥归来后，巴斯的名气更大了，从祖国的大江南北到世界各地的人都来看望巴斯。有一位在福州经商的台湾同胞，经常到大梦山看巴斯。先是他一人，后来他身边多了一个年轻人——他的儿子。起初他们看巴斯训练，觉得既新奇又不可思议。甚至巴斯在广州、深圳和厦门等地的巡回展出，他们也跟着去了。他们先是围着巴斯转，后来，那位年轻人的身影开始围着陈小玲"姐姐"转了。巴斯这才知道，他们看上了"姐姐"。

果然没多久，这位台湾商人找到陈玉村主任，央求他做媒，让"姐姐"嫁给他的儿子。但"姐姐"一直舍不得离开巴斯，直到1995年，眼看着自己已近30岁，这才洒泪离开了巴斯。

巴斯促成了这桩跨越海峡的良缘，而"姐姐"也一直不忘巴斯。随夫定居台湾的陈小玲"姐姐"仍时时关注着巴斯，不时回到大梦山来看望巴斯。只要她一呼唤巴斯，巴斯就能分辨出是她的声音，耳朵就会随着声音而转。

巴斯和陈小玲。
Basi and Chen Xiaoling.

A cross-strait marriage

In 1995, Chen Xiaoling, who had become a sister figure to Basi, reluctantly left after accompanying Basi for ten years and married to Taiwan.

As for Chen Xiaoling's marriage, Basi had been the matchmaker. In the ten years that Chen Xiaoling was with Basi, she took care of her and taught her sporting stunts. After returning from San Diego in the United States, Basi's reputation had frown. People from all over the country, from all corners of the world, came to Strait (Fuzhou) Panda World. There was a Taiwan businessman in Fuzhou who often went to Dameng Hill to watch Basi. He was alone before bringing a younger male—his son. They watched Basi train and were entranced. As Basi toured Guangzhou, Shenzhen and Xiamen, they followed. At first, they followed for Basi, but it would seem the young man began to follow Chen Xiaoling. Basi knew that this man was interested in her elder sister.

Sure enough, the businessman found Director Chen Yucun to play matchmaker between his son and Chen Xiaoling. However, Chen Xiaoling was reluctant to leave Basi, until 1995, seeing she was nearing the age of 30 she tearfully left Basi.

Basi contributed to this cross-strait marriage, and her sister Chen Xiaoling never forgot about the dear panda. Even after following her husband to Taiwan, Chen Xiaoling still kept up to date with news of Basi. From time to time, she would return to Dameng Hill to visit Basi. As soon as she called for Basi, the panda was able to recognize her voice and followed the sound until she found her former keeper.

巴斯和陈小玲。
Basi and Chen Xiaoling.

施飞宁"姐姐"走了

从1984年到1994年，陈小玲"姐姐"把她最美好的十年青春年华奉献给了巴斯。在这期间，邱梅霖、施飞宁两位"姐姐"也先后来到巴斯身边。1987年，巴斯和元元出访美国，就是陈小玲、邱梅霖两位"姐姐"陪她们去的，陈小玲照顾巴斯，邱梅霖照顾元元。再后来，邱梅霖到美国定居，但每次回国，她都会来看望巴斯。

三位"姐姐"就像跑接力赛一样，陪伴着巴斯走过"朝朝"，又迎来"暮暮"。最让巴斯难过的莫过于陪伴了她20多年的施飞宁"姐姐"在2016年突发脑部疾病而英年早逝。

1989年，18岁的施飞宁刚从学校毕业，就来到海峡（福州）熊猫世界，后来成了巴斯的专职饲养员。这不仅是她的第一份工作，也是她这辈子唯一的一份工作。当年跟她一起来到海峡（福州）熊猫世界的姑娘们，随着年龄的增长，陆续离开了这一工作岗位，只有她一如既往地坚守。

在施飞宁眼里，巴斯就像自己的孩子。有一次施飞宁的女儿来了，正好听到她在叫巴斯"宝贝"。她的女儿吃惊地问道："妈妈，你叫我'宝贝'，也叫巴斯'宝贝'，到底谁才是你的'宝贝'呀？"

尽管巴斯早已不是"小孩"，但总是喜欢黏着施飞宁。巴斯耍脾气不肯吃饭时，施飞宁也要哄要骗。施飞宁轮休时，换人喂巴斯，她就会推盆子不肯吃，即使喂进去了也会吐出来。为了巴斯，施飞宁不敢休假，在外头还得时时电话"遥控"。"带了巴斯这么多年，我真的把她当成自己的孩子了。"施飞宁曾说。

施飞宁给巴斯喂食。
Shi Feining was feeding Basi.

施飞宁为巴斯热敷。
Shi Feining was easing Basi's discomfort with a hot water bottle.

施飞宁为巴斯冲澡。
Shi Feining was bathing Basi.

The loss of a dear sister，Shi Feining

From 1984 to 1994，Chen Xiaoling dedicated a good decade of her youth to Basi，acting as a sister to the panda. During this period，two sisters，Qiu Meilin and Shi Feining，also came to Basi's side. In 1987，Basi and Yuanyuan visited the United States，accompanied by Chen Xiaoling and Qiu Meilin. Chen Xiaoling took care of Basi while Qiu Meilin looked after Yuanyuan. Later，Qiu Meilin would settle in the United States，but every time she returned to Fuzhou，she would visit Basi.

These three sister figures participated in a race accompanying and helping Basi to overcome the challenges she faced. It was therefore saddening for Basi when Shi Feining passed away from a brain disease in 2016.

Shi Feining came to Strait（Fuzhou）Panda World after graduating from school in 1989 at the age of 18. She would later become a full-time keeper for Basi. This was her first job，but also the only job she would have in her life. The girls who had entered Strait（Fuzhou）Panda World at the same time as Shi Feining gradually left their positions as they grew older，but Shi persisted.

In Shi Feining's eyes，Basi was like her own child. Once，Shi Feining's daughter came and heard her mother refer to Basi with an endearing Chinese term "*baobei*"，usually used for children and to mean treasure. The daughter asked her mum，"Mum，you call me and Basi your *baobei* but who is truly your *baobei*?"

Although Basi was no longer a cub，she always liked to stick with Shi Feining. When Basi threw tantrums and refused to eat，Shi Feining would try tricking Basi into eating. If Shi Feining was on a break and someone else tried to feed Basi，she would refuse to eat and swallow the whole bowl before spitting it all out. For Basi，Shi Feining rarely took vacations and was always reachable. "Being with Basi so many years，I treat her as if she is my own child，" Shi once said.

每天清晨6点多，施飞宁就来上班，晚上6点多才下班回家。如果遇到巴斯生病等突发情况，她就回不了家，加班加点也是常事。在施飞宁"姐姐"走的前几天，她还获得了"全国三八红旗手"称号。可惜证书都还没来得及看一眼，她竟突发脑溢血走了。默默呵护，精心照顾巴斯长达20多年的施飞宁，永远离开了她心爱的巴斯。从睡在大熊猫幼仔身边彻夜不离，悉心照顾，到专职照料巴斯，施飞宁将自己20多年的美好年华都献给了大熊猫们。

巴斯和施飞宁。
Basi and Shi Feining.

Every morning around 6 o'clock, Shi Feining came into work and would return home after 6 o'clock in the evening. If Basi was sick, it was the norm that she would not go home and work overtime. In the few days before Shi Feining passed away, she won an award—the national "March 8 Red-Banner Holder" on March 8th, the International Women's Day. Unfortunately, she suffered a brain haemorrhage before she was even able to look at her certificate. After over 20 years of looking after Basi, Shi Feining suddenly left her beloved panda. From sleeping with panda cubs at night and taking care of them to looking after Basi, Shi Feining dedicated a good twenty years of her life to taking care of giant pandas.

午后小憩　彧容叔德

巴斯精彩瞬间。
Wonderful moments of Basi's.

第四章 友谊与欢乐使者
Chapter 4 The ambassador of friendship and joy

在现场观看巴斯的美国观众。
The US audience were watching Basi.

1987年，巴斯和元元受原国家林业部派遣，代表中国野生动物保护协会赴美国圣地亚哥访问半年，轰动美国西海岸。

In 1987, Basi and Yuanyuan were sent to San Diego, the USA, to represent China Wildlife Conservation Association for half a year, causing a stir on the USA's west coast.

大熊猫旋风

　　1972年，美国前总统尼克松率团访华。这标志着自中华人民共和国成立后中美相互隔绝的局面终于打破。1979年1月1日，中美正式建立外交关系，从而结束了长达30年之久的不正常状态，中美关系进入了一段蜜月期。

　　在这样的背景下，1987年7月到1988年2月，受原国家林业部派遣，大熊猫巴斯、元元代表中国野生动物保护协会出访美国圣地亚哥市。

　　巴斯和元元乘飞机从中国飞往美国洛杉矶，再从洛杉矶坐卡车到圣地亚哥。她们乘坐的卡车外面挂着标语，上面写着"熊猫专车"，卡车还由当地警察一路护送。

　　虽然当时大熊猫在美国很受欢迎，但在巴斯和元元到来之前，圣地亚哥动物园里还从来没有大熊猫。所以巴斯、元元去的时候，当地人非常"疯狂"。由于来看大熊猫的人太多，每个人要排队5小时却只能观看3分钟。很多市民就一遍又一遍地排队，为的是能看清楚中国大熊猫。

美国民众排队等候观看巴斯。
The US people were waiting in line to view Basi.

　　1972年，美国前总统尼克松应周恩来总理的邀请首次访华。周恩来总理代表中国政府在答谢晚宴上正式宣布向美国赠送一对国宝大熊猫——原产于四川宝兴县的"玲玲"和"兴兴"。此后，中美之间关于大熊猫保护、疾病防治等方面的交流合作逐渐热络。

Taking San Diego by storm

In 1972, former US President Richard Nixon visited China. This marked the first instance of communication between the USA and China since the founding of the People's Republic of China in 1949. On the 1st January 1979, diplomatic relations were formally established, ending the awkward tension between China and the USA, and leading to a "honeymoon period" between the two countries.

It is in this political climate that Basi and Yuanyuan were sent to the USA to represent China Wildlife Conservation Association between July 1987 and February 1988.

Basi and Yuanyuan were flown from China to Los Angeles, where they would embark on a truck journey to San Diego. Their truck was painted with "Panda Truck" on the exterior and was escorted by the local police.

Although pandas were well liked by people in the USA, the San Diego Zoo had never hosted pandas until the arrival of Basi and Yuanyuan. As a result, when they arrived, the locals went crazy. People lined up for five hours just to see the Chinese pandas for three minutes, only to line up again.

巴斯"手"提公文包。
Basi was holding a briefcase.

At the invitation of former Premier Zhou Enlai, former US President Richard Nixon visited China for the first time in 1972. At dinner, Premier Zhou Enlai announced the gift of two giant pandas from China to the USA, Lingling and Xingxing from Sichuan. Since then, the two countries have cooperated frequently to protect giant pandas and prevent diseases for them.

巴斯为美国民众表演杠铃。
Basi was performing with barbells for the US audience.

本来展出双方约定巴斯访问的时间只有3个月。眼看3个月的期限到了，游客依然不减。在美国方面的再三请求下，经中国国务院批准后才把原定3个月的访问时间延长到了半年。

在巴斯访问圣地亚哥市的半年里，这座城市沸腾了。250万人次前往参观，轰动太平洋西海岸。每天购票观看的游人长龙达1千多米，和大熊猫有关的纪念品全部脱销。要知道，当时圣地亚哥市只有100万人口，250万人次的参观人潮从哪里来的呢？

时任圣地亚哥市长的奥康纳女士感叹道："中国的大熊猫来了之后，整个圣地亚哥市掉进了爱河！"

其他国家的媒体曾发出这样的报道："巴斯的这次出访，是中国大熊猫对外展览历史以来最为轰动、最为成功的一次。"

Originally, Basi was only meant to be in the USA for three months. However, as the three-month deadline approached, it was clear that interest in the pandas had not decreased as the number of visitors had not decreased at all. After repeated requests from the USA, the visit was extended to six months.

In the time that Basi spent in San Diego, she was visited by over 2.5 million people. The queue stretched over a kilometre, and every souvenir or panda-related merchandise was sold. This is considering that at the time, San Diego only had a population of one million! Ms. O'Connor, the mayor at the time exclaimed, "The city of San Diego has fallen in love with the pandas!" The visit received international media coverage, claiming that this visit was the most successful and sensational exhibit of China's giant pandas.

巴斯为美国观众献技。
Basi was performing for the US audience.

留根猫毛作纪念

天下没有不散的筵席。

转眼间，半年的时间很快就要过去了。眼看着巴斯和元元即将回国，圣地亚哥的许多市民舍不得和她们分别，有的市民还失声痛哭起来。

在她们离开的前一天晚上，一位老太太流着泪找到陪同她们出访的陈玉村："为了看巴斯，我已经第十次过来了。请您让我近距离地再看她一眼吧。"

陈玉村被老人恳切的请求感动，破例让她站在巴斯身边拍了一张照片。

圣地亚哥动物园总经理夫人凭着地利之便，一直守着巴斯。整整一个晚上，她都没有合眼。

凌晨4点，巴斯和元元分别乘车前往机场，沿途都是送别的人群。总经理夫人在陈玉村登机前拦住他，真诚地请求说："可不可以留一根巴斯的毛给我，让它陪伴我度过没有巴斯的岁月？"

陈玉村感动不已。他轻轻拍了拍巴斯的头说："老伙计，美国朋友要你一根毛，你答应不答应？"

巴斯扭头一看，正好看到她眼巴巴地盯着自己。巴斯的心软了，点了点头。陈玉村当即从巴斯身上轻轻拔下一根毛发，总经理夫人接过毛发，就放在嘴边亲吻了起来……

巴斯向美国观众道别。
Basi was bidding farewell to the US audience.

An unusual souvenir

All good things must come to an end.

In the blink of an eye，half a year had almost passed. Aware that Basi and Yuanyuan were to return home to China，many San Diegans were upset，with some even crying. On the evening before their departure，an elderly woman with tears in her eyes sought Chen Yucun，one of the staff who accompanied the pandas. He said，"I've come to see Basi ten times now. Please，let me get a bit closer."

Chen Yucun was touched by the sincerity of the elderly woman's request，and against protocol，allowed the woman to stand beside Basi and take a photo.

The San Diego Zoo general manager's wife too was upset，spending the whole night with Basi. At 4 a.m. as Basi and Yuanyuan were driven to the airport，there were people who had come to see them leave for along the way. As they arrived at the airport，the general manager's wife approached Chen Yucun，asking earnestly，"Would it be possible to leave me with one of Basi's hairs，so that I can remember her when she is gone?"

Chen Yucun，once again touched by the sentiment，gently asked Basi，"My dear friend，this American friend of ours would like one of your hairs. Would you be okay with that?"

Basi looked at the American，and seeing her emotional state，nodded her head. Chen Yucun then plucked a hair and gave it to the general manager's wife，who overcome with emotion，kissed it dearly.

巴斯在圣地亚哥动物园。
Basi was in the San Diego Zoo.

巴斯的故事还在美国流传

时任圣地亚哥动物园公关部职员的乔治妮，负责巴斯和元元在美国期间的接待和宣传工作，她先后接待了各国媒体采访。这些媒体刊发了各类报道20,000多篇，将巴斯的故事传遍全世界。巴斯因此得名"特技熊猫""友谊天使"。

乔治妮一直保存着巴斯和元元在美国时的照片和当时报道她们的报纸。1991年，乔治妮以巴斯在美国期间的故事为基础，创作出版了儿童图文书《两只熊猫的圣地亚哥动物园之旅》。她希望通过这本书告诉孩子们要保护大熊猫，爱护珍稀野生动物。乔治妮说："我觉得巴斯是友谊与和平的象征，她为中美交流搭起了一座桥梁。"

2015年11月28日，乔治妮应邀来到福州，她欣喜地看到巴斯还在继续书写着传奇故事。她穿着印有巴斯画像的套头衫，还带来印有巴斯的"中国红"长裤等纪念品。乔治妮告诉巴斯，她打算再写一本书，讲述这30多年来，发生在巴斯身上所有的奇迹和爱，让更多的人知道巴斯的传奇故事。

巴斯和乔治妮。
Basi and Georgeanne.

Basi's story lives on in the USA

Georgeanne, who was responsible for the promotion of Basi and Yuanyuan during their stay at the San Diego Zoo, received media interviews from different countries, leading to over 20,000 news reports about Basi. These reports spreaded Basi's story around the world, and Basi was known as an "angel of friendship".

报道巴斯和元元在美活动的报刊。
The magazine and newspapers reporting Basi and Yuanyuan's activities in the USA.

Georgeanne kept all the photos and news articles about Basi and Yuanyuan's visit. In 1991, using Basi's stay in the USA as a foundation, she wrote the children's book *The Visit of Two Giant Pandas at the San Diego Zoo* in order to inspire children to look after giant pandas, and beyond them, other wild animals. Speaking on Basi, she said, "I think that Basi is a symbol of friendship and peace. She has built a bridge on which China and the USA are able to communicate."

On 28th November 2015, Georgeanne was invited to Fuzhou. There, she was able to see Basi again, and was overjoyed. She wore a pullover and brought some souvenir such as socks that had Basi's images printed on them. Georgeanne was able to tell Basi that she intended to write another book about Basi and her experiences over the last thirty years.

巴斯和儿童图文书《两只熊猫的圣地亚哥动物园之旅》。
Basi and the children's picture book *The Visit of Two Giant Pandas at the San Diego Zoo*.

你從媽：處罰了嗎蝴蝶關愛的問候溫暖的一幕 璦容淑德

巴斯精彩瞬间。
Wonderful moments of Basi's.

第五章　熊猫"盼盼"与亚运雄风
Chapter 5　Panpan—the panda at the Asian Games

1990年，巴斯成为北京亚运会吉祥物"盼盼"的原型。
In 1990, Basi became the archetype of Panpan, the mascot for the Beijing Asian Games.

巴斯与"盼盼"

巴斯注定是个风云"人物"!

1990年9月22日至10月7日,第11届亚运会在北京举行。北京亚运会是中华人民共和国成立后第一次承办的大型国际综合性体育赛事。北京亚运会的举办在神州大地掀起了一股强劲的体育热潮。"我们亚洲,山是高昂的头……亚洲风乍起,亚洲雄风震天吼……"一首《亚运雄风》唱遍中国。北京亚运会迎来了中国体育事业前所未有的强盛,迎来了中国人民热爱体育运动前所未有的高潮。

在北京亚运会开幕式上,当吉祥物大熊猫"盼盼"出现在会场中央时,主持人激情四溢地说:"她召唤蓝天,召唤大地,召唤未来!"在北京亚运会举行期间,手持金牌做奔跑状的"盼盼"形象几乎天天出现在各大媒体上,同时也深深地印在了国人的脑海里。憨态可掬的亚运会吉祥物"盼盼",不仅征服了全国人民,同时也赢得了亚洲和世界人民的好感。

北京亚运会吉祥物盼盼。
Panpan, the mascot for the Beijing Asian Games.

1985年,在长春电影制片厂任美工的刘忠仁在报纸上看到了征集北京亚运会吉祥物设计作品的启事,他决定试一试。

北京亚运会筹委会在征稿时就明确规定用大熊猫做吉祥物。刘忠仁找来了一些大熊猫照片作为参考资料,说来也巧,其中就有几张是巴斯的照片。

看着巴斯做了很多模仿人的动作照片后,刘忠仁笑了。他暗暗地在心底做出决定:"巴斯,你就是我创作的原型了。"刘忠仁决定打破常规,创作一组拟人化的大熊猫图案,让它们不仅看上去可爱,还能体现北京亚运会"团结、友谊、进步"的主题。

In 1985, Liu Zhongren, an artist at Changchun Film Studio, saw an advertisement in the newspaper that requested designs for the Beijing Asian Games mascot. This piqued his interest and he decided to submit an entry.

The organising committee had clearly stated that a panda must be used for the mascot, so Liu Zhongren located some photos of pandas which he used as reference materials. It happened to be that some of these photos were of Basi.

After seeing that Basi imitated human actions, Liu Zhongren laughed. He made a decision at that point to use Basi as the archetype for his mascot. Liu Zhongren decided to do something unconventional—to draw a mascot that was anthropomorphic so as to make it look lovely and to reflect the theme of the Beijing Asian Games: unity, friendship and progress.

北京亚运会吉祥物盼盼。
Panpan, the mascot for the Beijing Asian Games.

Basi and Panpan

Basi was destined to be "a big card"!

The 11th Asian Games were held in Beijing from 22nd September to 7th October, 1990. This was to be the first large international sports event hosted by the People's Republic of China. China's role as the host lead to significant increases in participation in sports around the country. *The Spirit of Asia* could be heard across China, and this event resulted in the sports industry in China reaching a new high.

As Panpan appeared on the center of the venue, the host of the opening ceremony of the Beijing Asian Games declared, "She brings the blue sky, the earth, and the future!" Over the duration of the games, the image of Panpan running with a gold medal in its hand was a common sight. Panpan not only became a topic of interest and enthusiasm in China, but also won the favour of all Asia and around the world.

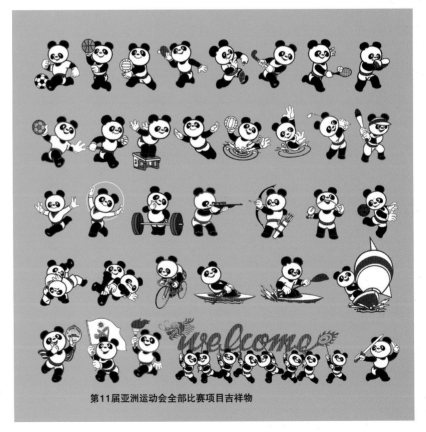

第11届亚洲运动会全部比赛项目吉祥物

在"盼盼"出现之前，历届亚运会从未有过专门的吉祥物，正是这只手持金牌的"盼盼"，成了中国吉祥物设计史上的里程碑。

有了吉祥物"盼盼"，还需要有"形象代言人"。这一光荣任务，就落到了大熊猫巴斯和青青的身上。她们从福州火速赶到北京。第二天，亚运会组委会召开新闻发布会说："亚运会吉祥物原型巴斯将会在亚运会的重要时刻出现，在另一个舞台上为来自亚洲各国的运动员'加油'助威。"

作为亚运会吉祥物"盼盼"的原型，巴斯的知名度很高，甚至超过了很多体育明星。那时候，北京亚运会冠军想和巴斯合影，也不是件容易的事，要经过审批才行。

Before the appearance of Panpan, there had not been a mascot designed for the Asian Games. As such, Panpan became a milestone in the history of Chinese mascots.

With the creation of Panpan came a role for Basi and Qingqing. They were rushed to Beijing from Fuzhou. The next day, the organising committee of the Beijing Asian Games held a press conference, announcing that "Basi will appear at a critical time in the Asian Games to cheer on the athletes."

Being the archetype for Panpan further alleviated Basi's fame and status, becoming more famous than many sports stars. At the time, it was only possible after an approval process for the gold medal winners of the Beijing Asian Games to take photos with Basi.

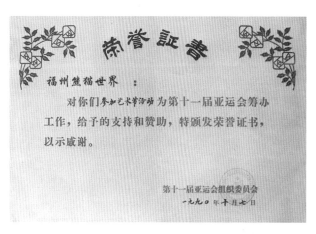

第11届亚运会组委会给福州熊猫世界颁发的荣誉证书。
The certificate of honour issued by the organising committee of the 11th Asian Games to Strait (Fuzhou) Panda World.

穿越天安门广场

1990年10月7日，北京亚运会闭幕当天中午，陈玉村主任突然接到亚运会组委会的通知："请巴斯赶到梅地亚中心参加记者招待会！"

陈玉村把巴斯请进一辆卡车，一路赶往梅地亚中心。快接近天安门广场时，交警把卡车拦了下来。

"卡车不能经过天安门广场！请绕行！"交警说。"车上是大熊猫巴斯，她要赶去参加记者招待会，没有时间绕道了。"陈玉村请求交警让道。

"大熊猫？开什么玩笑？"当车的后门打开，只见巴斯正坐在车里。见着交警愣了，巴斯伸出红红的舌头，开始向他卖起萌来。

"真是巴斯！"交警又惊又喜。经过请示上级，交警终于同意放行。载着巴斯的卡车顺利地通过了天安门广场。巴斯经过交警面前时，交警还向她敬了一个礼。

到了梅地亚中心，巴斯特别兴奋。巴斯从容登台表演，一口气把学会的体育项目几乎都表演了一遍，并挥动鲜花向大家问好。来自39个国家和地区报道亚运会的记者激动不已，纷纷从现场发回巴斯表演的报道。

在亚运会闭幕式当晚，巴斯的精彩表演让世界感受到中国体育事业的强国梦想，认识了"盼盼"这个代言中国体育走向世界的划时代符号。

巴斯进行体育运动。
Basi was doing sports.

Going through Tiananmen Square

On the 7th October 1990, the noon of the closing ceremony of the Beijing Asian Games, Chen Yucun received an urgent message from the games' organising committee, requesting Basi's presence at Media Centre for a press conference as soon as possible.

Ushering Basi into a small truck, Chen Yucun went to Media Centre in a hurry. However, upon reaching Tiananmen Square, they were stopped by a traffic policeman who informed them that no trucks were to cross Tiananmen Square, and that they should seek passage elsewhere.

Chen Yucun tried to explain the situation, telling the policeman that Basi was urgently needed for the press conference. The policeman replied, "A giant panda? Are you joking?" Upon opening the back of the truck, he was stunned to see a real panda there. Recognising the panda as Basi, the policeman was delighted. After consulting his superiors, Chen Yucun and Basi were released and they crossed Tiananmen Square. As Basi passed the policeman, he even saluted to her.

Upon arrival at Media Centre, Basi's excitement was clear. She got on stage and performed all the tricks she'd been taught, including waving flowers to greet everyone. The journalists from 39 countries and regions were ecstatic and delighted to report Basi's performances.

Basi's performance in the closing ceremony of the Asian Games expressed to the world China's dreams and aspirations in the areas of sports, and Panpan became a symbol for Chinese sports.

巴斯和元元。
Basi and Yuanyuan.

长盛不衰的"盼盼"风

北京亚运会结束后，"盼盼"的风头依然很盛。1990年，它的形象出现在中国城乡几乎每一个角落。许多人的背心上都印着大熊猫"盼盼"的图案。大熊猫"盼盼"造型的玩具也摆满了大小商店。由于"盼盼"的形象深入人心，以"盼盼"命名的防盗门、玩具、火柴等商品也风行神州大地。

The Panpan hype

After the conclusion of the Beijing Asian Games 1990，Panpan was still the topic of conversations. Panpan could be seen through various media forms in both rural and urban areas of China. Many people had Panpan shirts，Panpan toys and other forms of merchandise could be seen in many shops. As Panpan was so popular，many brands started using Panpan as advertising for a variety of goods ranging from toys to anti-theft doors.

巴斯和她的玩偶。
Basi and her doll.

春晚明星

春节是中国最重要的传统节日。1983年，首届现场直播形式的中央电视台春节联欢晚会（以下简称"春晚"）在除夕晚上播出，随后形成惯例。自此，央视"春晚"成为除夕之夜国人阖家团圆时不可或缺的一道"文化大餐"。

1991年2月14日，这天不仅是中国传统的除夕佳节，还是西方的情人节。就在这个特殊的日子，巴斯登上了万众瞩目的"春晚"舞台。

巴斯的节目安排在第11个，节目名称是《"盼盼"与巴斯》。

"去年，在北京有一件万众瞩目的大事，那就是第11届亚运会的召开。亚运的成就、亚运的光辉、亚运的精神至今还深深地印在每一位人的心里。尤其是亚运会的吉祥物大熊猫'盼盼'，更是受到了小朋友的喜爱。接下来，我们就来看看大熊猫'盼盼'的精彩表演！"

"各位观众，台湾同胞们，海外侨胞们，现在我们在中央电视台1,000平方米演播大厅现场直播超级巨星熊猫'盼盼'的无与伦比的表演。"

随着"春晚"主持人李扬、鞠萍热情洋溢的介绍，在观众雷鸣般的掌声中，巴斯迈着大熊猫特有的内八字步走上了舞台。

"这就是超级巨星大熊猫'盼盼'，她正走向篮球架，为大家做投篮表演。"投篮表演，四投三中，表现还不错。随后巴斯又来到杠铃场地。巴斯不清楚她练的是抓举动作还是挺举动作，面对杠铃，巴斯知道要举过头顶就算成功。巴斯力大无比，面对重重的杠铃，"举重若轻"。

看着巴斯轻轻松松就成功举起了杠铃，为巴斯配音的李扬开始使坏了。他骗巴斯说要一直举着，举够48小时才能得冠军。

巴斯举杠铃。
Basi was playing barbells.

Spring Festival celebrity

The Spring Festival, also known as Chinese New Year, is the most important of China's traditional festivals. In 1983, the first CCTV Spring Festival Gala was broadcast. Since then, CCTV Spring Festival Gala has become an indispensable part of Spring Festival celebrations for Chinese people on New Year's Eve.

The Spring Festival fell on the 14th February in 1991. This coincided with the Valentine's Day. On this particular day, Basi made an appearance on the Spring Festival Gala's stage. Her performance was the 11th of the night, titled "Basi and Panpan".

左图：巴斯举杠铃。
Left：Basi was playing barbells.

右图：巴斯在晃板上表演技艺。
Right：Basi was performing on the rocking band.

"Last year, Beijing hosted the 11th Asian Games. This was a significant event, and the achievements, the glory and the spirit, were felt deep in our heart. The mascot Panpan was loved by children. Now let us see the performance of our beloved Panpan!"

"Our dear fans at home, and our compatriots in Taiwan and overseas, we are live broadcasting the superstar Panpan's fantastic performance in the 1,000-square-meter CCTV studio!"

With this introduction from the hosts Li Yang and Ju Ping, and thunderous applause from the audience, Basi took to the stage.

"This is Panpan, the superstar panda. She is going to perform shooting hoops for us!" Basi was able to score three of the four shots, an amazing performance. After this, Basi proceeded to the weight station. Unsure whether she should snatch or jerk, she did know that she would pass the test if she lift the barbells above her head. And she was easily able to lift the heavy barbells.

Li Yang, watching her lift the barbells with ease, began to joke around, saying that she needed to hold the barbell for 48 hours to win the champion!

巴斯一听愣住了，精神有些恍惚。杠铃在她头顶上一阵晃动，随后砸在脑袋上，眼镜也掉在了地上。巴斯险些演砸了。巴斯静下心来仔细一想，"好像没有这样的规定吧。李扬肯定是骗我的。"幸亏巴斯从北京亚运会赛场归来不久，举重的比赛规则还是知道的。

想明白这一切后，巴斯再也不理唠唠叨叨的李扬，果断地放下杠铃，又赶紧奔向下一个赛场。巴斯才不上他的当呢！

为了缓解一下紧张的气氛，巴斯骑上自行车热身，绕着舞台转了好几圈。把筋骨活动开后，巴斯啃了几口竹子。补充完能量，巴斯开始了最激动人心的晃板表演。巴斯用后腿直立站在晃板上晃来荡去，"手上（前腿）"的功夫也不落下。先是套圈，巴斯一口气接住6个套圈，接着又进行飞叉旋转，挥舞得风生水起，让人眼花缭乱。

一番精彩表演下来，巴斯还意犹未尽，正思量着再表演什么项目时，主持人跑了过来，"巴斯的表演非常精彩，让我们为她颁奖！"巴斯获得了"春晚"历史上首个投篮、举重、骑自行车和晃板四项全能比赛冠军。

巴斯赶紧站在冠军领奖台上，让主持人把金牌挂在自己的脖子上，再接过奖杯和鲜花，最后面带微笑，向观众挥"手"致意。

巴斯玩晃板、套环。
Basi was playing on a rocking board with rings in her "hands".

巴斯站在领奖台上领奖杯。
Basi was standing on the podium with the trophy in her "hand".

Hearing this, Basi froze, lost as to what to do. In her shock, she even dropped her glasses after the barbell waggled over her head! She almost failed to complete the game. After a moment of thought, however, she quickly realised that this was impossible and Li Yang must have lied! Thanks to her time spent at the Asian Games, she knew the rules of weightlifting!

After thinking through all this, Basi put down the barbell and moved on to the next activity. Basi wasn't falling for Li Yang's tricks.

Basi proceeded to ride a bike around stage, then took a rest to eat some bamboo. Having replenished her energy, she was ready to take on the most difficult challenge. Standing on a rocking board on her hind legs, she used her front paws to catch six rings and perform more tricks.

After these brilliant performances, the host announced her to be the champion of hoop shooting, weightlifting, cycling and rocking board at the Spring Festival Gala. "Fantastic performance. Basi won the prize." the host said.

Basi stood proudly on the winners' podium, allowing the host to place a gold medal around her neck. Taking the trophy and flowers, she waved to the audience.

巴斯准备做卫生。
Basi wanted to "do some cleaning".

巴斯玩篮球。
Basi was playing basketball.

电话问候家乡亲人

其实，巴斯会表演的节目多着呢！巴斯不仅会表演投篮、举重、骑车、晃板等体育运动项目，她还会读书、看报，拉二胡，送公文等。

在"春晚"舞台上，巴斯想起了远在千里之外四川老家的父老乡亲，想起了《熊猫咪咪》这首歌。巴斯想告诉它们，她在福州生活得很好，还要问候它们，并对关心和照顾它们的人类朋友道一声谢谢。

巴斯环顾四周，发现了桌上的电话。她灵机一动，拿起了话筒："四川的亲人，你们好！我在福州生活得很好，请家乡父老乡亲放心。祝四川家乡人民新年快乐！"

登上"春晚"舞台后，巴斯更是名声大噪，成了家喻户晓的明星大熊猫。

巴斯给四川老家的父老乡亲打电话。
Basi was calling her relatives in Sichuan.

Telephoning the relatives at her hometown

The truth was, Basi was capable of many more performances. Other than the sports performance put on at the Spring Festival Gala, Basi was also capable of reading books and newspapers, playing the *erhu* and delivering documents.

At the Spring Festival Gala, Basi remembered her hometown in Sichuan, her fellow pandas, and the song *Panda Mimi*. She wanted to assure them that she was doing well and tell them that she enjoyed living in Fuzhou. She wanted to ask about their lives and pass on her thanks to the human beings who took care of them.

巴斯享用水果和饮料。
Basi was enjoying fruit and drinks, etc.

Looking around, Basi saw a phone on the nearby desk. She took the initiative, grabbing the phone, "My relatives in Sichuan, how are you? I'm doing very well in Fuzhou. Please be assured that I'm very happy. Wishing everyone at home a happy New Year!"

After performing on the Spring Festival Gala stage, Basi's fame only increased further as she embraced her role as a panda celebrity.

巴斯在读书。
Basi was reading.

大熊貓發現讓牠好奇的東西了起快奔向目的地到底有甚麼場景吸引着大熊貓

瑗容
淑德

巴斯精彩瞬间。
Wonderful moments of Basi's.

第六章 经历磨难的"坚强勇士"
Chapter 6 The warrior that has overcome hardships

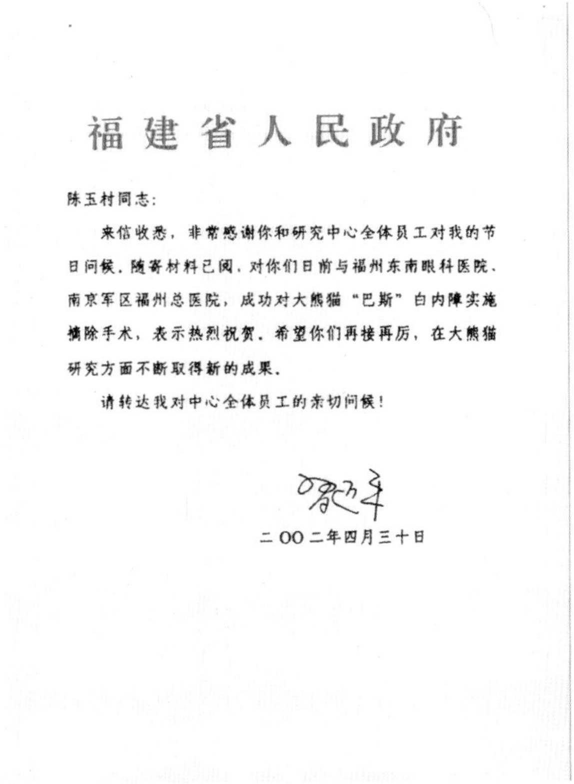

福建省人民政府

陈玉村同志：

　　来信收悉，非常感谢你和研究中心全体员工对我的节日问候。随寄材料已阅，对你们日前与福州东南眼科医院、南京军区福州总医院，成功对大熊猫"巴斯"白内障实施摘除手术，表示热烈祝贺。希望你们再接再厉，在大熊猫研究方面不断取得新的成果。

　　请转达我对中心全体员工的亲切问候！

二〇〇二年四月三十日

习近平对巴斯成功摘除白内障手术的贺信。
Xi Jinping's congratulatory letter on the successful surgery removing Basi's cataract.

　　2002年4月30日，巴斯白内障摘除手术成功后，时任福建省委副书记、省长的习近平向陈玉村发来贺信——"对你们日前与东南眼科医院、南京军区福州总医院，成功对大熊猫巴斯白内障实施摘除手术，表示热烈祝贺。希望你们再接再厉，在大熊猫研究方面不断取得新的成果。"

Basi underwent a successful cataract surgery on April 30th, 2002. Xi Jinping, then the deputy secretary of Fujian Provincial Committee of the Communist Party and governor of Fujian Province, sent a congratulatory letter to Chen Yucun— "I would like to express my warm congratulations on the successful removal of Basi's cataract by Fuzhou Southeast Eye Hospital and Fuzhou General Hospital of Nanjing Military Command. I hope you will continue to make great achievements in the giant panda research area."

海峡（福州）熊猫世界所在的大梦山与南京军区福州总医院仅相隔一条马路。这样的地利之便能保证巴斯和其他大熊猫身体出现问题时，南京军区福州总医院的专家们在10分钟内快速到达，及时救治。

Strait (Fuzhou) Panda World is located on Dameng Hill, one road away from Fuzhou General Hospital of Nanjing Military Command. This location is fortunate as the medical team from the hospital will arrive in 10 minutes in case any of the giant pandas required medical attention.

三次重生

回顾巴斯一生，她经历了三次灾难又重获新生，还经历了一次失明又重见光明。

第一次灾难是巴斯掉入冰河！是李兴玉"妈妈"等人救了她。

第二次是因高血压导致血管破裂，大出血而昏迷在鼓岭熊猫山庄！

第三次是得了肠炎并发急性胰腺炎而生命垂危！

第二和第三次是陈玉村主任和巴斯医疗团队拯救了她。

Basi's three times of rebirth

Looking back on Basi's life, she had three near-death experiences but was able to overcome them all. She suffered a bout of blindness and was able to regain her sight.

The first event was when Basi fell into the icy river and it was "mother" Li Xingyu who saved her.

The second time was a blood vessel rupture that resulted from high blood pressure. Basi fell into a coma at the summer resort in Kuliang because of heavy bleeding.

The third time was an onset of enteritis with acute pancreatitis putting her life in grave danger.

Basi was saved from the second and third incidents by Chen Yucun and Basi's medical team.

巴斯的医疗专家团队。
Basi's medical team.

2001年初，陈玉村在巴斯的笼舍内发现了血脚印，根据同类对比初步确定，巴斯罹患了高血压。这是全世界第一只确诊患有高血压的大熊猫。陈玉村在无奈之下，试着给巴斯服用降压药。幸运的是，巴斯服药后血压逐渐恢复平稳。

2002年4月，巴斯因患白内障被送往福州东南眼科医院。赵广健院长及其医疗团队联合南京军区福州总医院专家经过一个多小时的努力，成功对巴斯进行了国内外首例大熊猫白内障手术。

东南眼科医院赵广健院长在手术成功后表示："给大熊猫做白内障手术的过程与人一样，只不过眼球结构不同，最困难的是配合和术后护理问题。从目前情况看，手术比较顺利，巴斯的视力有所改善，为今后施行类似手术积累了经验。"

医疗专家们为巴斯进行体检。
The medical experts were conducting physical examinations on Basi.

At the beginning of 2001, Chen Yucun found bloody footprints in Basi's enclosure. After careful examinations and research, Chen Yucun realised that Basi was suffering from high blood pressure. This was the first case of a giant panda being diagnosed with hypertension. Chen Yucun couldn't help but try giving her anti-hypertensive medication. Luckily, Basi's blood pressure gradually returned to normal after the treatment.

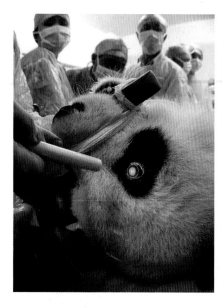

医疗专家检查巴斯的眼镜。
The medical experts were checking on Basi's eyes.

医疗专家准备为巴斯做手术。
The medical experts were preparing the operation on Basi.

In April 2002, Basi developed a cataract and was sent to Fuzhou Southeast Eye Hospital. The hospital director, Zhao Guangjian, and his medical team worked together with experts from Fuzhou General Hospital of Nanjing Military Command to operate on Basi. After more than an hour of surgery, Basi became the first giant panda success story for cataract removal worldwide.

After the operation, Zhao Guangjian said, "The process of cataract surgery for giant pandas is similar to that of humans, except that the eye structures are different. The most challenging aspect is team coordination and post-operative care. As of now, the surgery went relatively smooth. Basi's vision has improved and we have accumulated experience for similar future situations."

回到海峡（福州）熊猫世界后，科研人员和饲养人员采取了双人制日夜守护，每小时给巴斯滴上眼药水，并对她进行抗感染药物肌肉注射。经过近半个月的昼夜护理，巴斯终于重见光明。巴斯成了国内外首例手术摘除白内障成功的大熊猫。

　　2002年7月，在福州鼓岭熊猫山庄度夏的巴斯血压高出正常值两倍，鼻黏膜血管破裂造成大出血，昏迷了一周。也是在医疗团队专家们的抢救下，巴斯才最终挺了过来。

　　2010年6月，巴斯得了肠炎并发急性胰腺炎。一得知巴斯生病，陈玉村主任立即召集有关专家。专家们成立了抢救小组并马上对巴斯进行抽血，化验检测……他们发现巴斯的白细胞升高了两倍。巴斯四肢乏力，精神很差。经过三天的抢救，巴斯的健康状况还在不断恶化。陈玉村急切而果断地提出必须寻找新的治疗方案。

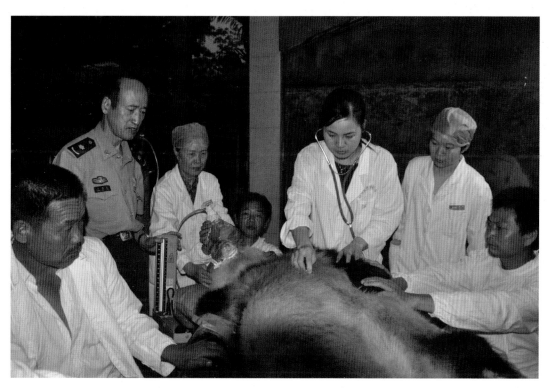

医疗专家为巴斯检查心跳，测量血压。
The medical experts were checking Basi's heart rate and blood pressure.

After returning to Strait (Fuzhou) Panda World, both researchers and keepers took care of Basi day and night. She was given eye drops every hour and her anti-infective medication was given intramuscularly. After nearly half a month of nursing, Basi finally regained her sight, making her the first giant panda to successfully have cataracts removed.

In July 2002, when Basi was spending the summer at the summer resort in Kuliang, her blood pressure rose to three times its normal rate. Due to this, her blood vessels in the nasal mucosa ruptured causing a massive haemorrhage. Basi was in a coma for a week but with a great medical team of experts, Basi regained consciousness.

Basi experienced another health issue in June 2010. She suddenly developed an onset of enteritis with acute pancreatitis. Upon learning that Basi was sick, Chen Yucun immediately called the experts together. The experts formed a rescue team in a timely manner to conduct a blood test on Basi. The results of this test showed that Basi's number of white blood cells had tripled. Basi's limbs lost their stiffness and she was in very low spirit. After three day's rescuing effort, Basi's health was still deteriorating. For fear of losing Basi, Chen Yucun was determined to ask for new treatments to help her.

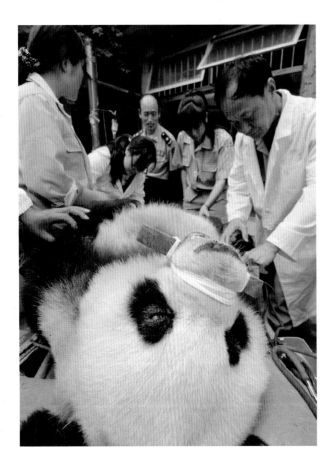

医疗专家为巴斯进行检查。
The medical experts were examining Basi.

抢救期间，陈玉村发现巴斯呻吟着无法入睡，头部大都向腹内弯曲，腿不能自然伸直，巴斯拉的黄色稀粪中还夹杂着其他颜色的黏液和泡沫，有患上胰腺炎的可能。但检验大熊猫患胰腺炎的指标，在当时的同行中还是个空白。尽管有不少人怕误治危及巴斯生命，陈玉村主任凭着30多年照顾大熊猫的经验，坚持立即让巴斯按并发胰腺炎的抢救方案进行治疗。

两天后，巴斯好转了不少，加上采用不麻醉的输液，巴斯又活过来了！

三次"死里逃生"，既是巴斯的顽强精神创造了生命奇迹，也是人类的不懈努力成就了这一生命辉煌。

后来，巴斯又患上了癫痫等疾病，还曾出现房颤、腰椎间盘突出、胸腔积液和腹水等症状，体力下降得厉害，走几步就喘粗气。巴斯的牙齿也渐渐退化，只能吃切成细块的竹笋。陈玉村和他的同事每天都会抽时间陪陪巴斯，给她按摩。饲养员每天喂巴斯特制的凉茶、蛋白粉、赤小豆、枸杞、红枣等，给巴斯调理身子。正是有了成功养护巴斯的先例，后来其他大熊猫的类似发病，只要症状一样，就能马上对症用药。

通过巴斯与疾病的抗争，人类进一步探索了大熊猫的养老问题。巴斯与疾病抗争的经历为老年大熊猫保护探索出了一整套行之有效的经验。

中科院动物研究所和海峡（福州）熊猫世界合作，通过将巴斯的体细胞植入去核后的兔子卵细胞中，在世界上最早克隆出一批大熊猫的早期胚胎。这一创新成果被评选为1999年中国十大科技进展。

海峡（福州）熊猫世界的工作人员彻夜照顾生病的巴斯。
The staff of Strait (Fuzhou) Panda World took care of Basi through the night when she was ill.

During this rescue effort, Chen Yucun found that Basi groaned and could not get to sleep. Her head was often slumped forwards and she could not straighten her hind legs. Basi's faeces were yellowy in colour mixed with other shades of mucus and foam which indicated that Basi may have pancreatitis. However, the symptoms of pancreatitis in giant pandas had not been discovered yet so this could not be certain. Those around Basi were worried about misdiagnoses and did not want to further jeopardise Basi's life, but Chen Yucun, who had more than 30 years of experience in taking care of giant pandas, insisted that Basi should undergo treatment for the pancreatitis as soon as possible.

After two days of treatment, Basi's health improved a lot and she seemed to be on the road to recovery. Together with the use of non-anaesthetic infusions, Basi was able to overcome this dangerous period in her health.

Basi's three narrow escapes from death were not only because of her strong spirit to survive, but also because of the persevering efforts of the humans around her. This, is the miracle of life.

海峡（福州）熊猫世界的工作人员为巴斯输液，喂药。
The staff of Strait (Fuzhou) Panda World were administering an infusion and medication for Basi.

Later, Basi would suffer bouts of epilepsy and other diseases. He had symptoms such as atrial fibrillation, lumbar disc herniation, pleural effusion, and ascites. Her physical strength deteriorated, and she began to breathe heavily after just a few steps. Basi's teeth had also gradually worn down over the years and she could only eat sliced bamboo shoots. Chen Yucun and his colleagues would find time every day to spend with Basi and to give her a massage. Daily, the keeper fed Basi nutritious foods such as protein powder, red beans, goji berries and red dates. It is because of their experiences of taking care of Basi, that humans are able to diagnose and treat other giant pandas when they begin to suffer similar health problems as Basi.

Basi's struggles with different health issues provided humans with a key to helping those aged pandas. It has given researchers more insight on how to prolong the lives of this wonderful species.

The researchers of the Institute of Zoology of the Chinese Academy of Sciences and Strait (Fuzhou) Panda World collaborated to implant Basi's somatic cells into enucleated rabbit egg cells and clone early embryos of giant panda's for the first time in the world. This scientific achievement was selected as one of China's top ten technological advances in 1999.

大熊貓喜歡聽音樂 拾是自己七嘴試吹 尺八不但能夠自得 其樂更是期待能夠 自愉媒扰分享愉悅 暖家淑德

巴斯精彩瞬间。
Wonderful moments of Basi's.

第七章　快乐长寿的"熊猫女王"

Chapter 7　Basi，the happy and long–lived Panda Queen

巴斯在37岁生日庆典上吃蛋糕。
Basi was eating the birthday cake on her 37th birthday celebration.

2017年，巴斯度过37岁生日，成为当时世界现存圈养最长寿的大熊猫，相当于人类100多岁的高龄。

Basi turned the age of 37 in 2017，becoming the world's oldest panda kept in captivity．That is equivalent to the human age of over 100 years old.

从2005年到2017年，海峡（福州）熊猫世界先后四次为巴斯举办生日庆典活动。每一次庆典，都让巴斯融入欢乐的海洋中。巴斯深深地感受到人类对她的情和爱。

李兴玉因抢救巴斯获原国家林业部颁发奖状。
Li Xingyu was awarded a certificate by the former State Ministry of Forestry for rescuing Basi.

2005年·巴斯见到了"妈妈"

2005年11月18日至12月18日，为庆贺巴斯25岁生日，"中国·福州第二届熊猫文化节"在福州举行。一支来自四川省宝兴县的巴斯"亲友团"应邀参加活动，"妈妈"李兴玉就在"亲友团"中。

巴斯的生日庆典在12月18日。一大早，巴斯刚走到花园，一个久远而又熟悉的声音传了过来："巴斯，妈妈看你来了！"李兴玉给巴斯带来了家乡的山苹果、鲜竹叶和山泉水。巴斯二话没说，从"妈妈"手中接过鲜竹叶、山苹果吃了起来。随后，巴斯手持红玫瑰，绕笼舍一周，向"妈妈"和老家亲友致意。

李兴玉"妈妈"默默垂泪："巴斯，你老了，我也老了……"站在一旁的陈玉村伤感地说："也许以后再也不会有这样一只聪明听话的大熊猫了！"

海峡（福州）熊猫世界将"巴斯救主"这一荣誉称号授予李兴玉。这是世界上独一无二的称号，巴斯为"妈妈"感到自豪和高兴。

From 2005 until 2017，Strait（Fuzhou）Panda World held four birthday celebrations for Basi. Every celebration made Basi overjoyed and she felt deeply loved by humans.

2005：Basi met her "mother"

To celebrate Basi's 25th birthday in 2005，the second China（Fuzhou）Panda Culture Festival was held in Fuzhou from November 18th to December 18th. A group of Basi's "relatives and friends" from Baoxing County，Sichuan was invited to participate in the event，among them was "mother" Li Xingyu.

Basi's 25th birthday celebration was held on the 18th of December. Early in the morning，Basi walked to the garden and could hear a familiar voice，"Basi，Mom is here to see you!" Li Xingyu brought Basi apples, fresh bamboo leaves and mountain spring water from their home region. Without another word，Basi took the food from her "mother" and ate. Afterwards，Basi held a bouquet of red roses and traversed around her enclosure to welcome her human "relatives" who had arrived to celebrate with her.

Li Xingyu silently shed tears, "Basi，you've grown old and so have I...." Chen Yucun，who stood beside Li Xingyu sadly commented，"Perhaps there will not be such a clever and obedient panda again."

Strait（Fuzhou）Panda World granted Li Xingyu the unique honorary title of Basi's Saviour. Basi felt a sense of pride for her "mother".

巴斯吃生日蛋糕。
Basi was eating birthday cake.

2010年·巴斯祝贺第16届广州亚运会召开

作为第11届北京亚运会吉祥物"盼盼"的原型，巴斯在她30岁生日之际，又一次与亚运会重逢，高龄巴斯因为亚运会再次焕发出青春的光彩。

2010年11月12日16时，随着中央电视台第16届广州亚运会开幕倒计时全球直播特别节目——《20年亚运跨越》将镜头切至国内唯一设置直播分场的海峡（福州）熊猫世界，主持人白岩松问道："大家还记得第11届北京亚运会吉祥物是什么吗？其原型是哪只大熊猫？这只大熊猫现在又在哪里？"此时直播镜头转向了海峡（福州）熊猫世界，巴斯再次出现在全球观众面前。

巴斯一会儿戴着眼镜看看电视，等候着亚运会开幕式的到来，一会儿又抱起广州亚运会吉祥物"乐羊羊"。此时，电视里正播放着北京亚运会吉祥物"盼盼"写给广州亚运吉祥物"乐羊羊"的一封信：

"20载亚运在我们两代吉祥物之间跨越，圣火依旧闪亮。祝愿广州亚运会圆满成功！……"

2010：Basi honoured the 16th Guangzhou Asian Games

中央电视台记者在海峡（福州）熊猫世界会场做直播。
A CCTV reporter was broadcasting live from Strait（Fuzhou）Panda World.

As the archetype of the 11th Beijing Asian Games mascot Panpan, on her 30th birthday, Basi was reunited with the Asian Games and experienced her youth once again.

At four o'clock in the afternoon on November 12th, 2010, the special program of countdown to opening of the 16th Asian Games in Guangzhou "The 20-year Asian Games Leap" was live broadcasted globally by CCTV. When the camera cut to Strait（Fuzhou）Panda World where the live broadcast was set, the host Bai Yansong asked, "Do you still remember what the mascot of the 11th Beijing Asian Games was? Which panda was the archetype? Where is the giant panda now?" And the camera panned to Strait（Fuzhou）Panda World, allowing Basi to appear in front of a global audience again.

One moment, Basi wore glasses for a while as she waited for the opening ceremony of the Asian Games to begin on the television, and the next she held on to the Guangzhou Asian Games mascots, Le Yangyang. At this time, the TV was broadcasting a letter from the Beijing Asian Games mascot Panpan to the current mascots Le Yangyang, "The 20-year Asian Games cross between our two generations of mascots. The sacred flame is still glowing. I wish the Guangzhou Asian Games a complete success!"

2015年·汇聚天下"有缘人"

"有缘人，巴斯喊你回福州。"2015年11月28日，在巴斯35岁生日庆典活动上，来自海内外的"巴斯有缘人"汇聚到了巴斯身边。

李兴玉、崔学振、赖炳辉、郑国芳、苏皖、赵广健、陈小玲、邱梅霖、施飞宁、乔治妮、刘忠仁、杨勇进、张云楷、卢维燮等来自中国乃至美国、曾经为巴斯献出爱心的人士以及南京军区福州总医院等团队获得了巴斯勋章。

这是海峡（福州）熊猫世界首次为爱心人士颁授巴斯勋章。在一枚枚巴斯勋章的背后，有着一个个感人的巴斯故事。勋章的获得者们都曾经历和见证了巴斯成长道路上的重要"人生节点"。每一个故事串起了巴斯传奇的成长历程。他们共同谱写出了一曲跨越地域、横跨世纪、跨越物种的"大爱之歌"。正是他们的爱心和善举，才让巴斯的生命意义非凡。

画家瑗容淑德所绘的巴斯主题画。
Paintings of Basi by a painter named Yuanrong Shude.

1991年就与巴斯结下缘分的台湾画家瑗容淑德2014年再次来到福州，待了50多天近距离观察巴斯的生活，之后花了1年多的时间创作了40幅以巴斯为主题的熊猫画。在巴斯35岁生日之际，瑗容淑德带着这些画作来福州为巴斯庆生。这些熊猫画，把巴斯从幼年、中年到老年的一生都形象地描绘了出来。"美丽的大熊猫巴斯，惬意地趴在湖边……到底梦见了什么呢？嘴角满满笑意。""大熊猫耍酷戴眼罩，还偷偷地移开，可是被发现了，害羞地低着头半遮着眼睛偷看。"和别的水墨画不同，瑗容淑德在每一幅熊猫画的旁边，都配了一句或动人或俏皮可爱的图说，就像给孩子讲故事一样。"把所有画串起来，就是一本好看的熊猫绘本故事书了。"瑗容淑德说。

2015: An affinitive gathering of the world

"Basi would like those whom she shares an affinity with to come back to Fuzhou." On 28th November 2015, Basi's 35th birthday celebration brought her friends from home and abroad to her side.

Li Xingyu, Cui Xuezhen, Lai Binghui, Zheng Guofang, Su Wan, Zhao Guangjian, Chen Xiaoling, Qiu Meilin, Shi Feining, Georgeanne Irvine, Liu Zhongren, Yang Yongjin, Zhang Yunkai, Lu Weixie and others from China, the United States and Fuzhou General Hospital of Nanjing Military Command reassembled to be presented with the Basi Medals.

This was Strait (Fuzhou) Panda World's inaugural medal awarding moment to the people who shared their love for Basi.

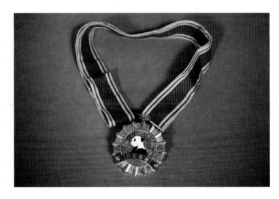

巴斯勋章。
A Basi Medal.

Each medal given to the recipient represented a touching story he had shared with Basi. They had witnessed and experienced important moments in Basi's life and each person contributed to Basi's story. Together, they wrote a song "The Great Love" that spans regions, centuries and species. It was their love and kindness that made Basi's life that much more extraordinary.

A Taiwanese painter named Yuanrong Shude who had the pleasure of meeting Basi in 1991 was once again reunited with Basi when she came to Fuzhou in 2014. She spent more than 50 days closely observing Basi, and spent more than a year drawing 40 paintings of the giant panda. On Basi's 35th birthday, Yuanrong Shude took

乔治妮在海峡（福州）熊猫世界接受记者采访。
Georgeanne was being interviewed by the reporters at Strait (Fuzhou) Panda World.

these paintings to Fuzhou to celebrate for Basi. The paintings portray Basi's life from childhood, to adolescence, to old age. "The beautiful giant panda, Basi, happily lingered by the lake... what could she be day dreaming about with that smile?" "She wears a pair of fashionable sunglasses and quietly moves away. When discovered, she is shy, bowing her head and lowering her eyelids halfway." Unlike other Chinese ink paintings, Yuanrong Shude had incorporated a moving line in each picture as if they were telling a children's story. "Stick all the pictures together and it becomes a story book about a panda."

2017年·最长寿大熊猫

2017年1月18日，海峡（福州）熊猫世界为巴斯举行37岁贺岁庆典。台湾省社会科学研究院院长孙武彦特地带来了3万盆台湾蝴蝶兰，将巴斯的贺岁庆典活动现场装扮得花团锦簇。

四川省大熊猫生态与文化建设促进会会长罗光泽给巴斯带来了珍贵的生日礼物——一是四川大熊猫画家童昌信创作的生动还原当年李兴玉跳入冰河抢救巴斯瞬间的画作《抢救巴斯图》，二是书法家司徒华创作的《功勋熊猫 传奇巴斯》书法作品。

海峡（福州）熊猫世界还为爱心人士颁发了第二批巴斯勋章。长期跟踪报道巴斯的《雅安日报》策划总监高富华、海峡（福州）大熊猫研究交流中心修云芳和徐素慧等人获颁巴斯勋章。

更让人兴奋的是，在贺岁庆典当天，巴斯的名字命名的"世界的巴斯 和平的图腾"纪念碑在海峡（福州）熊猫世界揭幕。该纪念碑由福州市人民政府、新华网和福州保护大熊猫协会共同设立。

在贺岁庆典上，巴斯以37岁的高龄，获得了世界纪录认证有限公司颁发的《世界现存最长寿圈养大熊猫》证书。巴斯离世前一周，还获得吉尼斯世界纪录有限公司颁发的《世界现存最长寿圈养大熊猫》证书。

At the celebration on January 18th, 2017, Basi was 37 years old and received the World Record Certification documenting her existence as the world's oldest living panda in captivity. And a week before Basi's departure, she received the Guinness World Record Certification as well.

2017: The longest living giant panda

Basi's 37th birthday celebration was held on January 18th, 2017 by Strait (Fuzhou) Panda World. Sun Wuyan, Dean of Taiwan Provincial Academy of Social Sciences, specially brought 30,000 Taiwanese moth orchids to adorn Basi's birthday celebration.

Luo Guangze, Chairman of Sichuan Giant Panda Ecology and Culture Promotion Association, brought two precious birthday gifts for Basi. The first was a painting "Rescuing Basi" by Sichuan panda painter Tong Changxin depicting Basi's fall into the icy river and being saved by Li Xingyu. The second present was a calligraphy work by renowned calligraphy master Situ Hua named "Feats of a Panda, the Legendary Basi" in the beautiful black ink.

Strait (Fuzhou) Panda World also granted a second round of Basi Medals. Gao Fuhua, the planning director of *Ya'an Daily*, who had long tracked and reported on Basi, and Xiu Yunfang, Xu Suhui of Strait (Fuzhou) Panda World were a few of those recipients.

Even more exciting, on the day of the ceremony, a monument commemorating Basi was unveiled at Strait (Fuzhou) Panda World.

巴斯吃生日蛋糕。
Basi was eating birthday cake.

为巴斯庆生的巴斯迷们。
Basi's fans celebrated her 37th birthday.

動物園運動大賽閉幕大熊貓巴斯美女壓倒性勝利獲得十項全能的冠軍得到金牌與獎盃得此殊榮全體熊貓擊掌慶賀

媛蓉淑德

巴斯精彩瞬间。
Wonderful moments of Basi's.

第八章　永恒的爱与和平
Chapter 8　Eternal love and peace

2017年9月13日，巴斯因病抢救无效离世。
On September 13th，2017，Basi passed away due to illness.

巴斯离世

有时现实是无情的，岁月不饶"人"，2017年9月13日，年事已高的巴斯走到了生命的尽头。创造了一个又一个难以复制的传奇故事的巴斯，还是离开了我们！

2017年9月14日上午，海峡（福州）熊猫世界召开新闻发布会。陈玉村主任在发布会上哽咽着宣布：第11届亚运会吉祥物"盼盼"的原型，国内外著名的友谊天使、大熊猫明星——巴斯，因病于2017年9月13日上午8时50分离世，享年37岁。

"巴斯于1984年5月6日来到福州，整整在这里生活了33个年头。在这期间，她得到了包括习近平总书记在内的多位国家领导人的关心和祝贺……巴斯曾为国家外事活动、科普教育、科学研究、两岸交流，以及为国家大熊猫栖息地保护工程筹集资金做出了卓越贡献。她不愧是一只传奇的大熊猫，世界的巴斯、和平的图腾。"陈玉村主任在发布会上深情地回忆道。

年老的巴斯。
Old Basi.

Basi's departure from the world

Sometimes it is difficult to accept reality. Life does not spare humans and neither does it spare Basi who with her old age, passed away on September 13th, 2017. After years of creating legendary memories for everyone, Basi had to leave us.

On the morning of September 14th, 2017, Strait (Fuzhou) Panda World held a press conference. Director Chen Yucun tearfully announced, "The archetype of the 11th Asian Games mascot Panpan, a famous angel of friendship, a giant panda star locally and abroad, Basi passed on at 8:50 on the morning of September 13th, 2017 at the age of 37."

医护人员精心照料巴斯。
The medical staff took care of Basi attentively.

Chen Yucun affectionately recalled all that Basi had contributed in her life time, "Basi came to Fuzhou in 1984 on May 6th and resided here for 33 years. During this period, she received care and well wishes from national leaders including Xi Jinping, General Secretary of CPC Central Committee. She made contributions to the country's foreign affairs, popular science education, scientific research and cross-strait relations. She also contributed to raising funds for the national giant panda habitat protection project. She was a legendary creature of the world and a representative of peace."

海峡（福州）熊猫世界的工作人员与巴斯告别。
Staff at Strait (Fuzhou) Panda World were saying their last goodbyes to Basi.

　　从2017年6月开始，巴斯的健康状况就不太好。医护人员检查后发现巴斯已有肝硬化、四肢水肿、鼻血不止、严重贫血等情况。4名饲养员每天24小时值班看守。巴斯的住所安装了6个高清摄像头，24小时监控记录她的一举一动。3个多月里饲养员和医护人员夜以继日地护理。巴斯最终仍因年老体衰、肝脏硬化、肾功能衰竭而离世。

　　陈玉村主任回忆道："巴斯生命的最后时刻是在2017年9月13日早上。我6点起来，先给她喂了水，再测量她的体温。体温还维持在37.2℃，但量了她的呼吸、心跳后吓了我一跳，次数都增加了一倍。"

　　巴斯的主管饲养员罗伟铭回忆说："2017年9月13日，她像往常一样，趴在床板上喘着气睡觉。突然我发觉她没有了呼吸声，心跳也十分微弱，慢慢地心跳也停了。她安静地离开了我们，永远地离开了！"

　　"巴斯走了，永远地离开我们了！不管你愿不愿意，她还是那么安详地走了，好像睡着了一样！我们动了动她，似乎还有点不相信。她就那么静静地趴着，没有痛苦。生命时钟永远定格在8时50分。我们打来热水，轻轻地为她擦拭眼眶、鼻子、毛发……虽然她经历了这最后100天病痛的折磨，但除了消瘦，她的毛发依旧黑白分明，依旧美丽。步入老年以后，她每天清晨都会在那棵棕榈树下坐着，累了就改趴着，静静地晒着太阳，仿佛在等着大家来上班，而我也养成每天看她一眼才能安心去工作的习惯。"海峡（福州）大熊猫交流研究中心徐素慧说道。

From June 2017, Basi's health had been deteriorating. After examinations, medical staff found that Basi had cirrhosis, severe oedema, nosebleeds and anaemia. Four keepers were on duty 24 hours a day to take care of Basi. Basi's enclosure was installed with 6 high-definition cameras to monitor her every move, every moment of the day and night. Over the last three months, keepers and medical staff had been nursing Basi around the clock. Basi eventually passed away due to her age, liver cirrhosis and renal failure.

Thinking back to that morning, Chen Yucun recounted that in Basi's final moments. Chen Yucun got up at 6 in the morning to feed her some water first and then measured her body temperature which had remained at 37.2 degrees Celsius. Her breathing and heartbeat however, scared Chen Yucun as it had doubled in frequency.

Basi's chief keeper, Luo Weiming recalled, "On September 13th, 2017, Basi lay on her bed as usual and panted. Suddenly, I realised that she had stopped breathing. Her heartbeat was fading and it slowly came to a stop. She left us quietly and forever."

Xu Suhui, a researcher of Strait (Fuzhou) Panda World, also gave a few words, "Regardless of whether you like it or not, Basi has left us peacefully as if she were just asleep. We tried to move her and could not believe it. She was so peaceful. There was no pain as her life clock stopped and would forever stay at 8：50. We got some warm water and gently wiped her eyelids, nose, and hair. Although she endured a lot through her illness for the last 100 days and experienced some weight loss, her fur was still a distinct black and white, eternally beautiful. During her later years, she would sit under the palm tree every morning. If she was tired, she would lie on her stomach and quietly bask under the sun, waiting for everyone to come to work. I too, developed a habit of having a look at her before starting my work every day."

大熊猫巴斯。
Basi, the giant panda.

依依送别

大梦山麓，棕榈青竹。只是再也见不到巴斯晒太阳、吃竹子的画面了。

在巴斯去世的第三天（2017年9月16日）上午10时，巴斯悼念会在海峡（福州）熊猫世界举行。1,000多名巴斯迷赶来与她道别。

巴斯迷遍及全球，前来参加送别的就有从美国、日本赶来的，也有从比利时、新加坡等国发来信件表达哀思的。巴斯去世的消息公布后，海峡（福州）熊猫世界两天内收到全球1,000多封来信。

巴斯曾经历三次生死难关均奇迹生还，但这一次，她真的离开了！多希望看到她能再调皮地向我们吐个舌头……巴斯躺在冰柜里，比平日瘦了些，毛色依然黑白分明、干净美丽。

来自日本的巴斯迷千秋吉富赶来送巴斯最后一程。她抱着巴斯造型的玩偶，用不太流利的中文告诉大家：

巴斯迷们写下对巴斯的祝福。
Basi fans wrote down their good wishes for Basi.

前来为巴斯送别的巴斯迷。
Fans who came to say goodbye to Basi.

"我喜欢巴斯，我是巴斯迷。"

一位50多岁的巴斯迷在得知巴斯去世的消息后，马不停蹄地从美国乘飞机回来，"我要送巴斯最后一程。"

悼念会现场留言板上，写满了各地巴斯迷的寄语。

"巴斯在弥留之际，毛色还是那么漂亮。"陈玉村边说边把花瓣撒向巴斯生活过的地方。这里四处悬挂着她各个年龄段的照片。人们隔着玻璃墙，向巴斯致哀献花，久久不愿离去。

A farewell

No longer will there be a chance again to see Basi basking in the sun and eating bamboo. On September 16th, the third day of Basi's death, a memorial was held at 10 o'clock in the morning at Strait（Fuzhou）Panda World.

More than 1,000 fans came to say goodbye to her from around the world. Some came from the United States and Japan to bid her farewell, while letters were sent from Belgium, Singapore and other countries to express their sorrow and condolences. After releasing the news of Basi's death, Strait（Fuzhou）Panda World received more than 1,000 letters from around the world within two days.

Having had three near-death experiences and still miraculously survived, this was the final time Basi would leave us. We very much would like to see her playfully stick her tongue out at us again... people saw Basi lying in the freezer, thinner than usual, and her coat was still black and white, clean and beautiful.

Chiba Gigi from Japan came to Basi to say goodbye. She was holding a Basi doll and told everyone in broken Chinese that "I like Basi, and I am a Basi fan."

When a 50-year-old Basi fan heard of the news, he rushed over the United States without stopping. "I am going to send Basi on this final leg of her journey."

On the message board for Basi at the memorial, there were many notes from her fans.

"Even until she left us, Basi's fur was still pretty." said Director Chen Yucun as he scattered petals over what once was Basi's enclosure. Hanging around the enclosure were pictures of Basi at all ages. People stayed for a long time as they continued to bring flowers in for their beloved panda.

前来为巴斯送别的巴斯迷。
Fans who came to say goodbye to Basi.

仿真巴斯

一个生物巴斯走了，另一个文化巴斯来了。

巴斯以她传奇的成长经历、国宝级的文化形象和世界范围内的关注度成为中国熊猫文化的标志，成为中国文化元素和中国文化品牌具有代表性的宝贵资源。

为了满足全球巴斯迷爱好大熊猫，关心巴斯的心愿，福州保护大熊猫协会正在筹建巴斯博物馆和巴斯俱乐部。中国科学院自动化信息研究所艺术中心主任张之益正在开展以巴斯生物基因结合科技创新的仿真巴斯科研工程，通过高新科技让人们在多维空间共享巴斯带来的文化魅力。张之益及其团队基于中国科学院的科技优势，依托前沿科技应用，将全方位再现巴斯的动作行为和体态特征。通过交互程序，以动静结合、人机互动的趣味体验方式，让体验者能"亲临"巴斯传奇一生中重要的精彩瞬间。

仿真巴斯效果图。
An effect chart of simulated Basi.

Remembering Basi

A panda loved by all may have left this world，but her spirit lives on.

Basi has become a symbol of China's panda culture with her renowned growing experiences，national treasure-level cultural image and worldwide attention. Basi has become a valuable resource of China's cultural element and brands.

In order to satisfy wishes of the global fans who love pandas especially Basi，Fuzhou Conservation Giant Panda Association plans to build the Basi Museum and set up the Basi Club. Zhang Zhiyi, director of the Art Center of Institute of Automation of the Chinese Academy of Sciences is carrying out a scientific project to simulate Basi. It combines Basi's biological genetics with technological innovations to enable people to share the charm of Basi. Based on the scientific and technological advantages of the Chinese Academy of Sciences，Zhang Zhiyi and his team rely on cutting-edge scientific technology to mimic Basi's actions，behaviours and characteristics. Through an interactive program，the technological interaction enables people to experience a "visit" to the important moments of Basi's life.

大熊猫巴斯。
Basi, the giant panda.

美麗的
大熊貓
巴斯
盼盼
愜意
的臥
在湖
邊見
蝴蝶
輕輕
飛不
敢打
擾牠
到底
夢見
了什
麼呢
嘴角
滿滿
笑意

瑗容　淑德

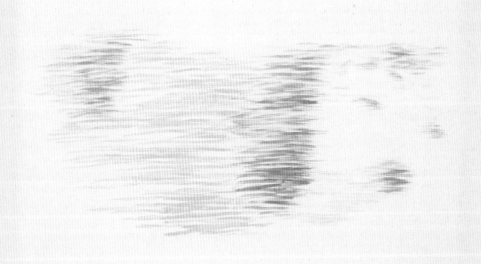

巴斯精彩瞬间。
Wonderful moments of Basi's.

尾声　巴斯小语
Endnote　Basi's biography

巴斯坐在树下晒太阳。
Basi was sunbathing under a tree.

我叫巴斯。我生于1980年，用现在的话说，我是"80后"。

我的老家在四川省宝兴县。记得那是1984年的春天，饥饿的我被困在"巴斯沟"的冰河上，生命危在旦夕。是李兴玉和她的侄儿冒险把我救起，并为我取名"巴斯"。数月后，福州成了我的第二个故乡。从此我有了父爱如山的熊猫"爸爸"陈玉村，有了温柔可亲的"姐姐"陈小玲、邱梅霖和施飞宁。我得到了无微不至的关怀，过着被人类宠爱的快乐生活。

1987年我第一次走出国门，才知道世界那么大，喜欢我的人那么多。我在美国圣地亚哥做了半年的亲善大使，交了250多万新朋友，凭我的才艺和技能轰动了美国西海岸。

没想到这才是我明星旅程的开始！1990年我受邀参加北京第11届亚运会，成了吉祥物"盼盼"的原型，成为传播亚运精神的象征。更为幸运的是，几个月后，彭丽媛"妈妈"来看我。在她的鼓励下，我登上央视"春晚"的舞台，过足了明星瘾。

时间又过了10年。我代表中国野生动物保护协会，到广州、深圳、珠海、济南等地，为实施大熊猫栖息地保护筹集资金！为国争光，为国宝出力，成为我生命的一部分。

巴斯和她的玩偶。
Basi and her doll.

Hi! My name is Basi. I was born in 1980, or as they say nowadays, I am an 80s baby.

My hometown is Baoxing County in Sichuan Province. I remember that in the spring of 1984, I was hungry and trapped in an icy river which was referred to as Basi Ditch. My life was in danger but it was Li Xingyu and her nephew who risked their lives and saved me before giving me my name. Three months later, Fuzhou became my second hometown. Since then, I was able to gain a "dad", Chen Yucun, and my caring "sisters", Chen Xiaoling, Qiu Meilin and Shi Feining. I received so much love and care, living a happy life.

巴斯在享受美味食物。
Basi was enjoying her delicious food.

I first went travelling outside of China in 1987. I learnt that the world was so much bigger and there were so many people who loved me. I worked as a goodwill ambassador for half a year in San Diego in the United States, and made more than 2.5 million new friends! The talents and skills I learnt in Fuzhou stirred the audience in America.

Little did I know that I would start on my journey of fame! In 1990, I was invited to participate in the 11th Asian Games held in Beijing, becoming the archetype behind the mascot Panpan, representing the spirit of the Asian Games. Even luckier, I was visited by Peng Liyuan. With her encouragement, I performed on CCTV's Spring Festival Gala stage! I became a superstar panda.

Another ten years passed and I represented China Wildlife Conservation Association, touring through places such as Guangzhou, Shenzhen, Zhuhai and Jinan to raise funds for a project that protected giant pandas' habitats. For the glory of my country, to help protect our national treasures, these events have become a part of my life.

海峡（福州）熊猫世界内的巴斯雕像。
The sculpture of Basi at Strait (Fuzhou) Panda World.

可是熊猫的岁月也不饶人啊，不知不觉我已老态龙钟！我和人类一样也得了白内障。是福州东南眼科医院、南京军区福州总医院的医生为我做了眼科手术。我居然成为国内外首只通过手术成功摘除白内障的大熊猫。

之后，我又成为世界上第一例确诊患有高血压的大熊猫，被送往千米高山上避暑疗养。由于激动，我血压上升导致血管破裂，血流不止，整整昏迷一周后再次起死回生。30岁生日前夕，我突然倒地不起，把所有关爱我的人都吓坏了。他们甚至为我安排了后事。但把我爱到骨子里的熊猫"爸爸"陈玉村，硬是和医生们一起把我从死亡线上给拽回来，让我再一次获得生命。

如今，我真的很老了！37岁的我相当于人类100多岁了，已经相当高寿了！虽然曾经的辉煌已成往事，但我知足了。有谁能得到比我更多的爱呢？可是时间都去哪儿了？眼看步子慢了，眼睛花了，竹子也啃不动了，但每天都有很多人来看我，有老朋友也有新朋友，连孩子们都知道我叫巴斯。

我活了这么大岁数，只想告诉大家一句话——这辈子能做一只大熊猫，值了！

——摘自《熊猫巴斯：时间都到哪儿去了》

37岁时的巴斯。
37 years old Basi.

Life however，does not spare pandas any more than it does to humans. Before I knew it，I had become old! I had cataracts like humans and the doctors at Fuzhou Southeast Eye Hospital and Fuzhou General Hospital of Nanjing Military Command performed an eye surgery for me. I actually became the first giant panda in the world to successfully undergo this procedure.

Not only that，I also became the world's first panda to be diagnosed with hypertension and was taken to the mountains away from the summer heat to recover. However，getting overly excited caused my blood pressure to rise and my blood vessels ruptured. The blood would not stop flowing and I was in a coma for a week before I awoke. On the eve of my 30th birthday，I suddenly toppled over and frightened those who loved me. They even prepared to arrange my funeral affairs but my "dad"，Chen Yucun，who loved me to the bones brought me back with the help of the doctors from the brink of death and gave me life once more.

Today，I am really old! I am 37 years old which is more than 100 years old for the mankind. Although the immense attention I received is now a thing of the past，I am satisfied. Who has received more love than me? But where has the time gone? Life is starting to slow before me and my eyes can no longer see what they used to，and bamboo is not as easy to chew as it once was. Still，many people come to visit me，friends old and new. Even children know that I am Basi.

In my long life，I want to tell you just one thing—the life of a giant panda is worth it!

—excerpt from *Basi the panda：where has the time gone*?

在2017全球大熊猫颁奖盛典大会上，巴斯获得终生成就奖。
Basi received the Lifetime Achievement Award at the 2017 Giant Panda Global Awards Ceremony.

（按拼音顺序排列）

编　　委：陈玉村　　高富华　　杨勇进　　周跃进

图片摄影：包　华　　陈小玲　　　陈玉村　　高富华　　高华康
　　　　　　刘可耕　　刘忠仁　　　吕　明　　乔治妮　　邱　军
　　　　　　王丽芳　　王晓峰　　　魏培金　　肖和勇　　杨婀娜
　　　　　　叶　诚　　瑗容淑德　　张丽君　　张之益　　张中琴
　　　　　　赵马峰　　祝敏松

图片提供：海峡（福州）熊猫世界　　新华网　　中新社

图书在版编目（CIP）数据

和平使者熊猫巴斯：汉英对照/高富华编著；
（新西兰）蒋梓青，（新西兰）蒋梓恒，陈思嘉译.--福州：
福建人民出版社，2018.9

ISBN 978-7-211-07982-7

Ⅰ.①和… Ⅱ.①高… ②蒋… ③蒋… ④陈…
Ⅲ.①大熊猫－普及读物－汉、英 Ⅳ.①Q959.838-49

中国版本图书馆CIP数据核字（2018）第186787号

和平使者熊猫巴斯
HEPING SHIZHE XIONGMAO BASI

作　　者：高富华
翻　　译：Donna Jiang（蒋梓青）　　Rocky Jiang（蒋梓恒）　　陈思嘉
责任编辑：周跃进
出版发行：福建人民出版社　　　　　　电　　话：0591-87533169（发行部）
网　　址：http://www.fjpph.com　　　电子邮箱：fjpph7211@126.com
地　　址：福州市东水路76号　　　　　邮　　编：350001
经　　销：福建新华发行（集团）有限责任公司
印　　刷：福州市东南彩色印务有限公司
地　　址：福州市金山浦上工业区冠浦路151号
开　　本：700毫米×1000毫米　1/16
印　　张：6.25
字　　数：160千字
版　　次：2018年9月第1版　　　　　　2018年9月第1次印刷
书　　号：ISBN 978-7-211-07982-7
定　　价：88.00元